Standards-Based Math

1-2

Written by
Vicky Shiotsu

Editor: Carla Hamaguchi

Illustrator: Darcy Tom

Designer/Production: Moonhee Pak/Cari Helstrom

Cover Designer: Barbara Peterson

Art Director: Tom Cochrane

Project Director: Carolea Williams

Table of Contents

Introduction

Each book in the *Power Practice*™ series contains dozens of ready-to-use activity pages to provide students with skill practice. The fun activities can be used to supplement and enhance what you are already teaching in your classroom. Give an activity page to students as independent class work, or send the pages home as homework to reinforce skills taught in class. An answer key is included for quick reference.

Standards-Based Math 1–2 provides activities that will directly assist students in practicing and reinforcing math skills such as place value, addition, subtraction, patterning, word problems, measurement, probability, and more! The book is based on the National Council of Teachers of Mathematics (NCTM) standards and divided into five main sections: Number and Operations, Algebra, Geometry, Measurement, and Data Analysis and Probability. You will find several activity pages in each section that will motivate students and reinforce their math skills.

Use these ready-to-go activities to "recharge" skill review and give students the power to succeed!

Name ______________________________ Date ______________

Batter Up!

Number and Operations

Write the answers on the baseballs.

A.

1 + 2 =   3 + 2 =   3 + 6 =

B.

4 + 2 =   3 + 0 =   2 + 6 =

C.

5 + 5 =   1 + 1 =   5 + 3 =

D.

3 + 4 =   7 + 2 =   2 + 5 =

E.

4 + 6 =   5 + 4 =   1 + 8 =

Hop and Add

Number and Operations

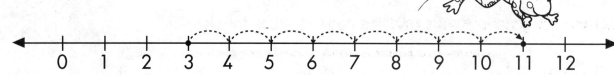

$$0 \quad 1 \quad 2 \quad 3 \quad 4 \quad 5 \quad 6 \quad 7 \quad 8 \quad 9 \quad 10 \quad 11 \quad 12$$

Use the number line to help you solve the problems.

To solve 3 + 8, start at 3. Then take 8 hops to the right to get your answer.

A. $5 + 5 =$ _____ $7 + 4 =$ _____

B. $1 + 8 =$ _____ $6 + 6 =$ _____ $9 + 3 =$ _____

C. $2 + 9 =$ _____ $3 + 7 =$ _____ $4 + 5 =$ _____

D. $6 + 5 =$ _____ $8 + 4 =$ _____ $1 + 9 =$ _____

E. $3 + 8 =$ _____ $7 + 5 =$ _____ $3 + 9 =$ _____

Colorful Flowers

Number and Operations

Add. Then use your answers and the code to color the flowers.

Code

13 – red	15 – yellow	17 – purple
14 – blue	16 – orange	18 – green

A.
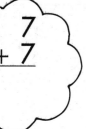

$$\begin{array}{r} 7 \\ + 7 \\ \hline \end{array}$$
$$\begin{array}{r} 4 \\ + 9 \\ \hline \end{array}$$
$$\begin{array}{r} 6 \\ + 8 \\ \hline \end{array}$$
$$\begin{array}{r} 8 \\ + 7 \\ \hline \end{array}$$

B.

$$\begin{array}{r} 9 \\ + 6 \\ \hline \end{array}$$
$$\begin{array}{r} 8 \\ + 8 \\ \hline \end{array}$$
$$\begin{array}{r} 7 \\ + 6 \\ \hline \end{array}$$
$$\begin{array}{r} 7 \\ + 9 \\ \hline \end{array}$$

C.

$$\begin{array}{r} 9 \\ + 8 \\ \hline \end{array}$$
$$\begin{array}{r} 6 \\ + 9 \\ \hline \end{array}$$
$$\begin{array}{r} 9 \\ + 5 \\ \hline \end{array}$$
$$\begin{array}{r} 8 \\ + 5 \\ \hline \end{array}$$

D.
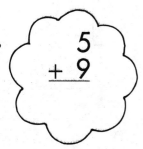

$$\begin{array}{r} 5 \\ + 9 \\ \hline \end{array}$$
$$\begin{array}{r} 9 \\ + 9 \\ \hline \end{array}$$
$$\begin{array}{r} 7 \\ + 8 \\ \hline \end{array}$$
$$\begin{array}{r} 9 \\ + 7 \\ \hline \end{array}$$

Number Detective

Number and Operations

Help find the missing numbers. Complete the facts below.

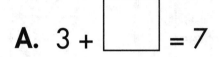

A. $3 + \boxed{} = 7$ $5 + \boxed{} = 10$

B. $\boxed{} + 6 = 11$ $\boxed{} + 9 = 12$ $\boxed{} + 7 = 9$

C. $8 + \boxed{} = 14$ $2 + \boxed{} = 11$ $7 + \boxed{} = 10$

D. $\boxed{} + 7 = 12$ $\boxed{} + 3 = 9$ $\boxed{} + 6 = 12$

E. $4 + \boxed{} = 11$ $9 + \boxed{} = 13$ $6 + \boxed{} = 13$

F. $1 + \boxed{} = 10$ $3 + \boxed{} = 11$ $5 + \boxed{} = 14$

G. $\boxed{} + 8 = 16$ $\boxed{} + 7 = 14$ $\boxed{} + 9 = 15$

H. $4 + \boxed{} = 12$ $8 + \boxed{} = 15$ $9 + \boxed{} = 17$

Standards-Based Math • 1–2 © 2004 Creative Teaching Press

Name _____ Date _____

Picnic Time

Number and Operations

Help the ant get to the picnic basket. First solve the problems. Then color the boxes that have odd numbers as answers to make a path.

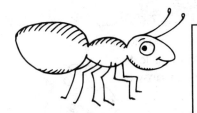

	$6 - 1 =$ ___	$5 - 4 =$ ___	$3 - 1 =$ ___
$5 - 3 =$ ___	$2 - 0 =$ ___	$7 - 4 =$ ___	$9 - 1 =$ ___
$8 - 4 =$ ___	$10 - 2 =$ ___	$8 - 3 =$ ___	$6 - 4 =$ ___
$10 - 5 =$ ___	$8 - 5 =$ ___	$9 - 4 =$ ___	$7 - 1 =$ ___
$6 - 3 =$ ___	$8 - 6 =$ ___	$7 - 3 =$ ___	$10 - 8 =$ ___
$9 - 2 =$ ___	$10 - 6 =$ ___	$8 - 2 =$ ___	$9 - 3 =$ ___
$10 - 3 =$ ___	$9 - 6 =$ ___	$10 - 1 =$ ___	

Standards-Based Math • 1-2 © 2004 Creative Teaching Press

Hop and Subtract

Number and Operations

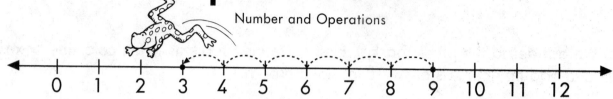

Use the number line to help you solve the problems.

To solve 9 – 6, start at 9.
Then take 6 hops to the
left to get your answer.

A. $8 - 4 =$ _____ $10 - 3 =$ _____

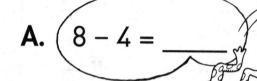

B. $9 - 7 =$ _____ $8 - 6 =$ _____ $11 - 8 =$ _____

C. $12 - 5 =$ _____ $8 - 2 =$ _____ $12 - 9 =$ _____

D. $12 - 4 =$ _____ $11 - 6 =$ _____ $11 - 3 =$ _____

E. $12 - 6 =$ _____ $10 - 8 =$ _____ $11 - 9 =$ _____

Standards-Based Math • 1–2 © 2004 Creative Teaching Press

Subtraction Roundup

Number and Operations

Subtract.

A.
12
− 7

13
− 6

12
− 4

13
− 8

11
− 5

B.
15
− 9

13
− 9

14
− 7

12
− 6

14
− 9

C.
13
− 5

15
− 6

13
− 4

16
− 9

14
− 6

D.
13
− 7

16
− 8

14
− 5

15
− 8

17
− 9

E.
18
− 9

15
− 7

17
− 8

14
− 8

16
− 7

Name _____ Date _____

Splish, Splash!

Number and Operations

Solve the problems. Write the answers in the raindrops.

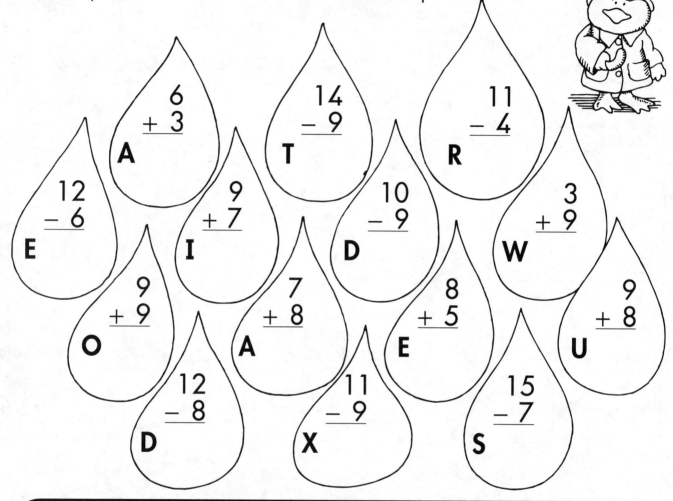

$$\begin{array}{r} 6 \\ + 3 \\ \hline \end{array}$$ A

$$\begin{array}{r} 14 \\ - 9 \\ \hline \end{array}$$ T

$$\begin{array}{r} 11 \\ - 4 \\ \hline \end{array}$$ R

$$\begin{array}{r} 12 \\ - 6 \\ \hline \end{array}$$ E

$$\begin{array}{r} 9 \\ + 7 \\ \hline \end{array}$$ I

$$\begin{array}{r} 10 \\ - 9 \\ \hline \end{array}$$ D

$$\begin{array}{r} 3 \\ + 9 \\ \hline \end{array}$$ W

$$\begin{array}{r} 9 \\ + 9 \\ \hline \end{array}$$ O

$$\begin{array}{r} 7 \\ + 8 \\ \hline \end{array}$$ A

$$\begin{array}{r} 8 \\ + 5 \\ \hline \end{array}$$ E

$$\begin{array}{r} 9 \\ + 8 \\ \hline \end{array}$$ U

$$\begin{array}{r} 12 \\ - 8 \\ \hline \end{array}$$ D

$$\begin{array}{r} 11 \\ - 9 \\ \hline \end{array}$$ X

$$\begin{array}{r} 15 \\ - 7 \\ \hline \end{array}$$ S

Riddle Time! What kind of suit does a duck wear?

To find out, first look at the answers you wrote.
Then write the letters that are beside them on the matching lines below.

Answer:

___ ___   ___ ___ ___ ___ ___
16   5   12   6   9   7   8

___ ___ ___ ___ ___ ___ ___ !
15   1   17   2   13   4   18

Standards-Based Math • 1–2 © 2004 Creative Teaching Press

Name _____ Date _____

Add It Up
Number and Operations

Miss Gray's students must read at least 6 hours to earn a free book. Look at the charts. Write the total number of hours the students read. Did they earn a book? Circle **Yes** or **No**.

Sandra	Hours
Week 1	1
Week 2	3
Week 3	2
Total	
Yes	**No**

Will	Hours
Week 1	1
Week 2	1
Week 3	1
Total	
Yes	**No**

Mackenzie	Hours
Week 1	1
Week 2	1
Week 3	2
Total	
Yes	**No**

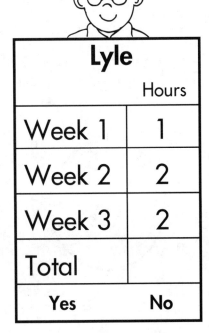

Sam	Hours
Week 1	2
Week 2	4
Week 3	2
Total	
Yes	**No**

Lyle	Hours
Week 1	1
Week 2	2
Week 3	2
Total	
Yes	**No**

Ashley	Hours
Week 1	2
Week 2	3
Week 3	2
Total	
Yes	**No**

Party Time

Number and Operations

Add. Write the answers in the balloons.

A.
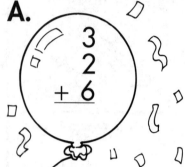
$$\begin{array}{r} 3 \\ 2 \\ +\ 6 \\ \hline \end{array}$$

B.
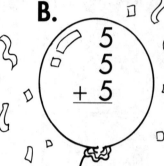
$$\begin{array}{r} 5 \\ 5 \\ +\ 5 \\ \hline \end{array}$$

C.
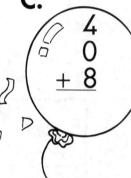
$$\begin{array}{r} 4 \\ 0 \\ +\ 8 \\ \hline \end{array}$$

D.

$$\begin{array}{r} 7 \\ 6 \\ +\ 1 \\ \hline \end{array}$$

E.
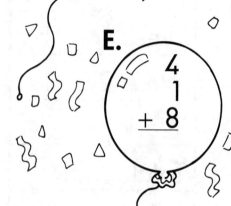
$$\begin{array}{r} 4 \\ 1 \\ +\ 8 \\ \hline \end{array}$$

F.

$$\begin{array}{r} 2 \\ 5 \\ +\ 4 \\ \hline \end{array}$$

G.
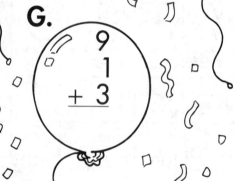
$$\begin{array}{r} 9 \\ 1 \\ +\ 3 \\ \hline \end{array}$$

H.

$$\begin{array}{r} 8 \\ 3 \\ +\ 3 \\ \hline \end{array}$$

I.

$$\begin{array}{r} 7 \\ 3 \\ +\ 5 \\ \hline \end{array}$$

J.
$$\begin{array}{r} 2 \\ 8 \\ +\ 8 \\ \hline \end{array}$$

K.
$$\begin{array}{r} 7 \\ 2 \\ +\ 7 \\ \hline \end{array}$$

L.
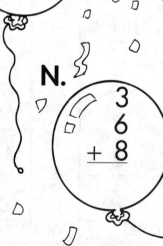
$$\begin{array}{r} 2 \\ 6 \\ +\ 5 \\ \hline \end{array}$$

M.
$$\begin{array}{r} 5 \\ 4 \\ +\ 9 \\ \hline \end{array}$$

N.
$$\begin{array}{r} 3 \\ 6 \\ +\ 8 \\ \hline \end{array}$$

Button Math

Number and Operations

A. Color 10 red.
   Color the rest yellow.

How many?

____ ten and ____ ones = ____

B. Color 10 blue.
   Color the rest purple.

How many?

____ ten and ____ ones = ____

C. Color 10 yellow.
   Color the rest green.

How many?

____ ten and ____ ones = ____

D. Color 10 orange.
   Color the rest red.

How many?

____ ten and ____ ones = ____

Name _____ Date _____

Bundles of Ten

Number and Operations

Write the number of tens. Then write the matching numbers.

A.

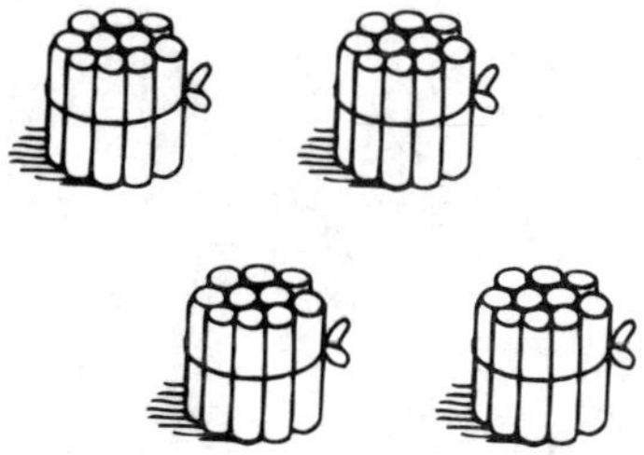

__4__ tens | 40

B.

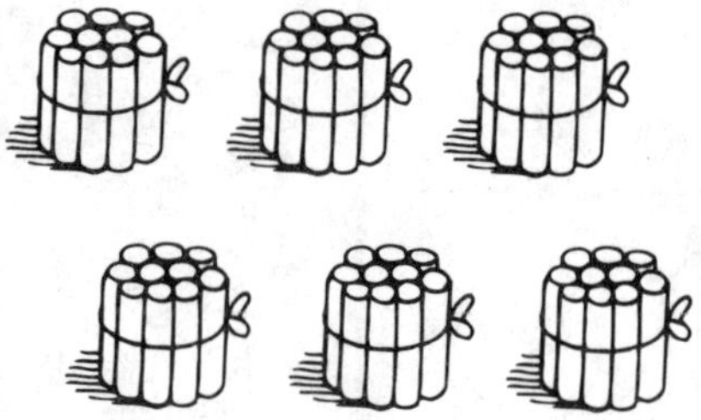

____ tens

C.

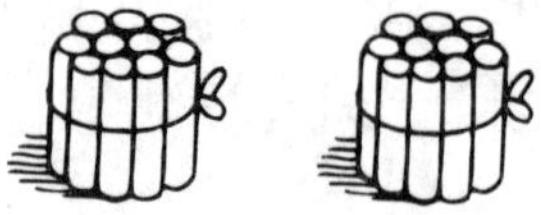

____ tens

D.

____ ten

E.

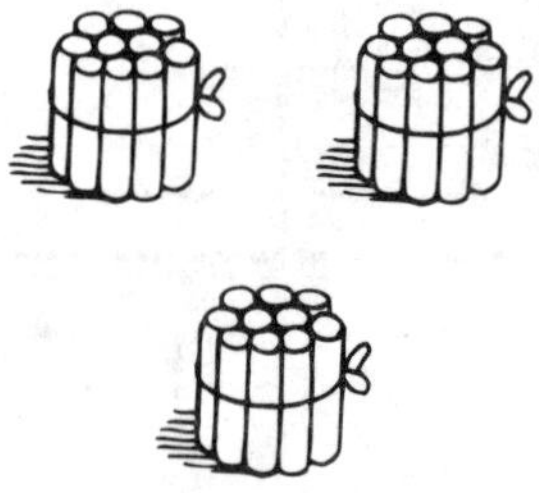

____ tens

F.

____ tens

G.

____ tens

H.

____ tens

I.

____ tens

Standards-Based Math • 1–2 © 2004 Creative Teaching Press

Name _____ Date _____

Tens and Ones

Number and Operations

Write how many tens and ones there are. Then write the number.

A.

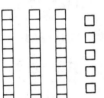

tens	ones
3	5

We write __35__.

B.

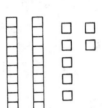

tens	ones

We write _____.

C.

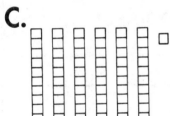

tens	ones

We write _____.

D.

tens	ones

We write _____.

E.

tens	ones

We write _____.

F.

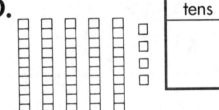

tens	ones

We write _____.

G.

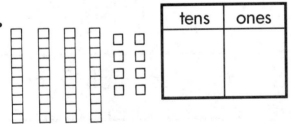

tens	ones

We write _____.

H.

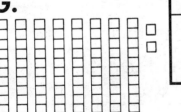

tens	ones

We write _____.

Number Riddles

Number and Operations

Solve the number riddles. Circle the correct answers.

A. I have 3 tens and 2 ones.
What number am I?

32   23

B. I have 1 ten and 6 ones.
What number am I?

61   16

C. I have 4 tens and
less than 5 ones.
What number am I?

47   43

D. I have 8 tens and
more than 7 ones.
What number am I?

89   86

E. I have 2 ones and
more than 6 tens.
What number am I?

72   52

F. I have 9 ones and
less than 4 tens.
What number am I?

49   29

G. I have more than 6 tens
and less than 5 ones.
What number am I?

33   91

H. I have less than 5 tens
and more than 4 ones.
What number am I?

54   45

I. I have more than 3 ones
and less than 7 tens.
What number am I?

88   56

J. I have less than 7 ones
and more than 5 tens.
What number am I?

55   61

Standards-Based Math • 1–2 © 2004 Creative Teaching Press

All Aboard!

Number and Operations

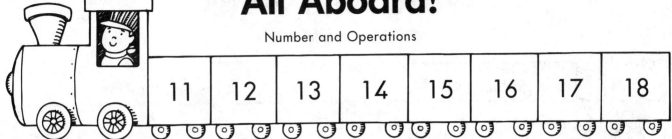

| 11 | 12 | 13 | 14 | 15 | 16 | 17 | 18 |

Write the missing numbers.

A. 18, 19, 20, _____, _____, _____, 24, _____, _____, 27

B. 34, 35, 36, _____, _____, _____, 40, _____, _____, 43

C. 56, 57, 58, _____, _____, _____, 62, _____, _____, 65

D. 65, 66, 67, _____, _____, _____, 71, _____, _____, 74

E. 88, 89, _____, _____, _____, 93, _____, _____, _____, 97

Write the numbers that come before and after.

F.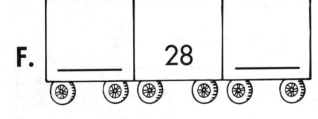

G.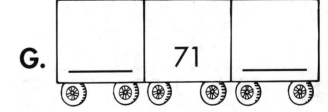

H.

I.

Watch the Signs

Number and Operations

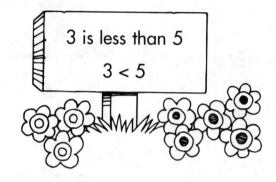

Write > or < in each ◯ .

A. 7 ◯ 2   3 ◯ 9   10 ◯ 8

B. 16 ◯ 18   23 ◯ 32   40 ◯ 30

C. 26 ◯ 20   9 ◯ 29   64 ◯ 48

D. 73 ◯ 55   44 ◯ 80   91 ◯ 19

E. 38 ◯ 83   51 ◯ 31   70 ◯ 90

F. 60 ◯ 56   87 ◯ 78   69 ◯ 72

Standards-Based Math • 1–2 © 2004 Creative Teaching Press

Name _____ Date _____

Rabbit's Riddle

Number and Operations

Rabbit has a riddle for you!

What do you call a rabbit who tells jokes?

To solve the riddle, first answer each problem below.
Then look at your answers and the letters in the boxes.
Write the letters on the matching lines at the bottom of the page.
(You will not use all the answers and letters.)

tens	ones
2	6
+	3
2	9

f 13 + 5	**b** 66 + 2	**n** 22 + 5	**n** 41 + 7
t 22 + 6	**n** 33 + 5	**y** 65 + 4	**u** 74 + 3
u 70 + 6	**r** 91 + 7	**n** 81 + 4	**y** 56 + 3

A __ __ __ __ __ __ __ __ __ __ __ !
 18 76 48 38 59   68 77 27 85 69

Starry Subtraction

Number and Operations

Subtract.

tens	ones
3	7
−	5
3	2

A.
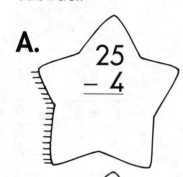

25
− 4

17
− 2

28
− 8

B.
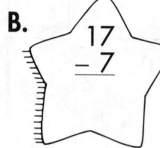

17
− 7

69
− 2

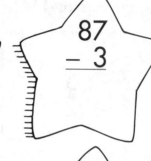

87
− 3

59
− 8

C.
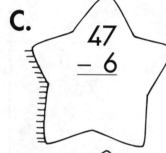

47
− 6

38
− 4

52
− 2

75
− 4

D.
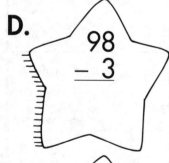

98
− 3

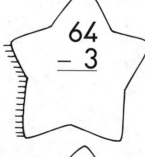

64
− 3

88
− 8

45
− 3

E.
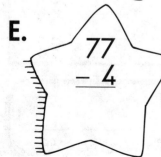

77
− 4

59
− 6

97
− 6

83
− 3

Subtracting One-Digit Numbers from Two-Digit Numbers without Regrouping

Standards-Based Math • 1–2 © 2004 Creative Teaching Press

Name _____ Date _____

In the Clouds

Number and Operations

Add or subtract.

	tens	ones		tens	ones
	5	2		3	7
+	1	6	−		5
	6	8		3	2

A.
23
+41

33
+30

B.
27
−14

56
−33

67
+11

12
+56

31
+47

C.
84
−12

32
−20

61
+14

64
−11

13
+16

D.
33
+30

18
+41

25
−20

85
−80

42
+24

E.
75
−65

61
+24

84
+15

30
+45

59
−22

Eager Beavers

Number and Operations

Add the ones.
Regroup the ones to make
  1 more ten.
Then add the tens.

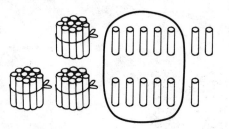

tens	ones
¹1	7
+ 2	6
4	3

A.

63	57	24	38	41
+18	+25	+29	+57	+19

B.

12	45	36	17	26
+19	+36	+59	+33	+ 9

C.

53	74	29	47	88
+17	+18	+14	+ 6	+ 3

D.

34	72	73	69	56
+59	+18	+ 9	+26	+24

Standards-Based Math • 1–2 © 2004 Creative Teaching Press

Counting Coins

Number and Operations

quarter
25 cents
25¢

half-dollar
50 cents
50¢

Count the coins. Write how many cents there are.

A.

_____ ¢

B.

_____ ¢

C.

_____ ¢

D.

_____ ¢

E.

_____ ¢

F.

_____ ¢

G.

_____ ¢

H.

_____ ¢

I.

_____ ¢

Let's Go Shopping

Number and Operations

Cross out the coins you will need to buy each item.

A. 15¢

B. 50¢

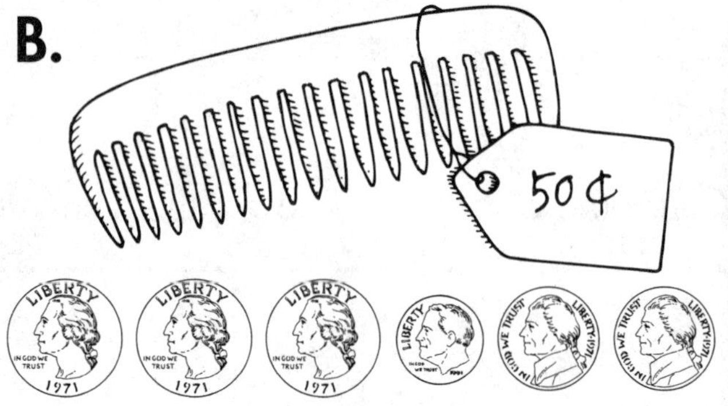

C. BULLDOG 40¢

D. 55¢

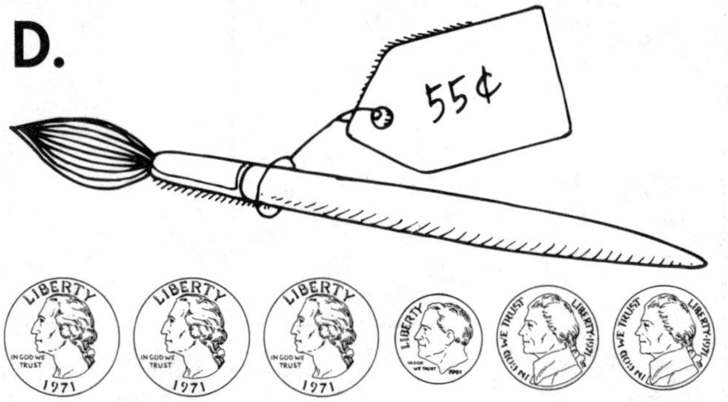

E. 75¢

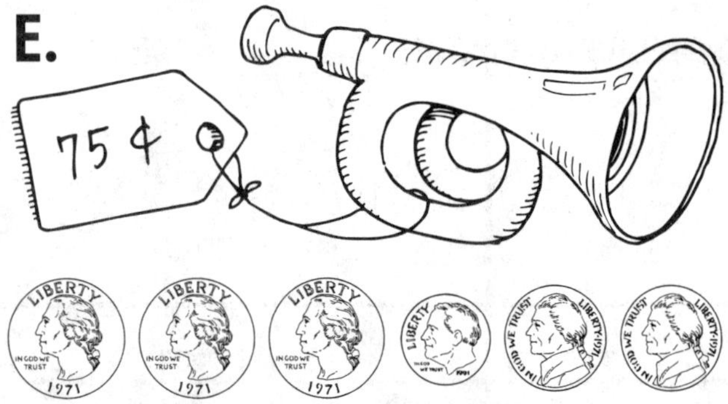

F. 80¢

G. 70¢

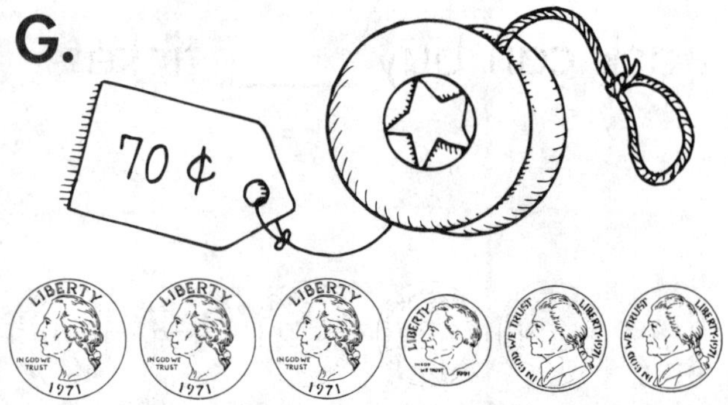

H. 90¢

Name _____ Date _____

Zoo Tickets

Number and Operations

Each zoo ticket costs 25¢.
Write the most number of tickets each person can buy.
Then cross out the coins each person needs to buy the tickets.

A.

Frank can buy _____ tickets.

He will have _____ ¢ left.

B.

Mary can buy _____ tickets.

She will have _____ ¢ left.

C.

Carla can buy _____ tickets.

She will have _____ ¢ left.

D.

Tom can buy _____ tickets.

He will have _____ ¢ left.

E.

Lily can buy _____ tickets.

She will have _____ ¢ left.

F.

Jack can buy _____ tickets.

He will have _____ ¢ left.

Name _____ Date _____

What Makes a Dollar?

Number and Operations

dollar
100 cents = $1.00

A. How many pennies make a dollar? _____

B. How many nickels make a dollar? _____

C. How many dimes make a dollar? _____

D. How many quarters make a dollar? _____

E. How many half-dollars make a dollar? _____

Draw two other coin combinations that make a dollar.
An example has been done for you.

25¢ 25¢ 5¢ 25¢ 10¢ 10¢		

How Much Money?

Number and Operations

Write how much money is in each set.

A.

_____

B.

_____

C.

_____

D.

_____

E.

_____

F.

_____

G.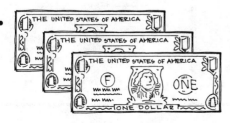

_____

Standards-Based Math • 1–2 © 2004 Creative Teaching Press

Find the Fraction

Number and Operations

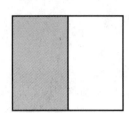

$\dfrac{1}{2}$ ← tells how many parts you are looking at

← tells how many equal parts there are in all

Circle the fraction that shows what part is shaded.

A.

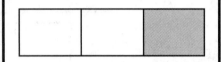

$\dfrac{1}{2}$ $\dfrac{1}{3}$ $\dfrac{1}{4}$

B.

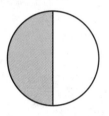

$\dfrac{1}{2}$ $\dfrac{1}{3}$ $\dfrac{1}{4}$

C.

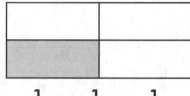

$\dfrac{1}{2}$ $\dfrac{1}{3}$ $\dfrac{1}{4}$

D.

$\dfrac{1}{3}$ $\dfrac{1}{5}$ $\dfrac{1}{6}$

E.

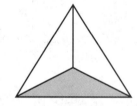

$\dfrac{1}{3}$ $\dfrac{1}{5}$ $\dfrac{1}{6}$

F.

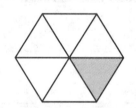

$\dfrac{1}{3}$ $\dfrac{1}{5}$ $\dfrac{1}{6}$

G.

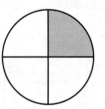

$\dfrac{1}{4}$ $\dfrac{1}{6}$ $\dfrac{1}{8}$

H.

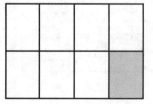

$\dfrac{1}{4}$ $\dfrac{1}{6}$ $\dfrac{1}{8}$

I.

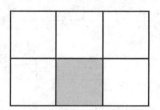

$\dfrac{1}{4}$ $\dfrac{1}{6}$ $\dfrac{1}{8}$

A Closer Look at Fractions

Number and Operations

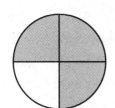

$\frac{3}{4}$ ← tells how many parts you are looking at

← tells how many equal parts there are in all

Write the fraction that tells what part is shaded.

A. 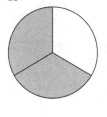	**B.** 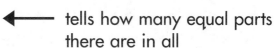	**C.** 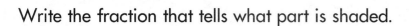
D. 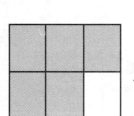	**E.** 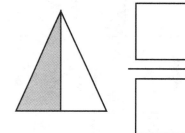	**F.** 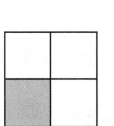
G. 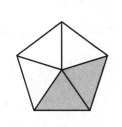	**H.** 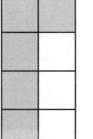	**I.** 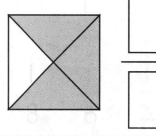

Equal Groups

Number and Operations

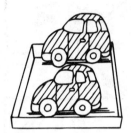 

$$\frac{1}{2}$$ ← tells how many parts you are looking at

← tells how many equal parts there are in all

Look at each fraction. Color the pictures to match.

A.

$$\frac{1}{4}$$

B.

$$\frac{1}{3}$$

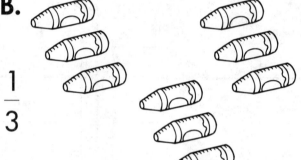

C.

$$\frac{1}{2}$$

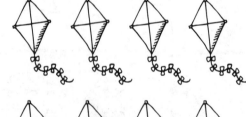
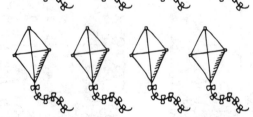

D.

$$\frac{1}{6}$$

 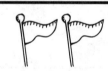
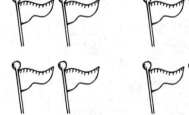

E.

$$\frac{1}{3}$$

F.

$$\frac{1}{4}$$

Standards-Based Math • 1–2 © 2004 Creative Teaching Press

Name _____   Date _____

Fishy Fractions

Number and Operations

Circle the fish to match each fraction.
Color the fish you circled.

$\dfrac{3}{4}$

A.

$\dfrac{2}{3}$

B.

$\dfrac{2}{5}$

C.

$\dfrac{3}{4}$

D.

$\dfrac{1}{2}$

E.

$\dfrac{3}{6}$

F.

$\dfrac{1}{3}$

G.

$\dfrac{1}{2}$

H.

$\dfrac{4}{5}$

Leafy Multiplication

Number and Operations

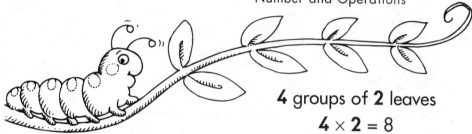

4 groups of **2** leaves
$4 \times 2 = 8$

Multiply. Use the leaves at the right to help you.

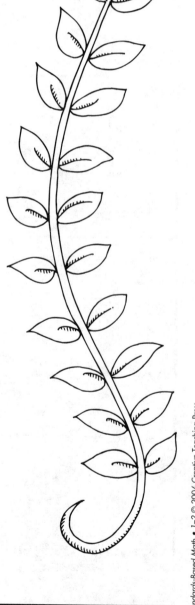

A. $5 \times 2 =$ _____

B. $1 \times 2 =$ _____

C. $8 \times 2 =$ _____

D. $3 \times 2 =$ _____

E. $10 \times 2 =$ _____

F. $11 \times 2 =$ _____

G. $2 \times 2 =$ _____

H. $4 \times 2 =$ _____

I. $7 \times 2 =$ _____

J. $6 \times 2 =$ _____

K. $12 \times 2 =$ _____

L. $9 \times 2 =$ _____

List your answers in order from the least to greatest.

_____, _____, _____, _____, _____, _____,

_____, _____, _____, _____, _____, _____

What pattern do you see?

_____

Standards-Based Math • 1–2 © 2004 Creative Teaching Press

Pretty Petals

Number and Operations

3 groups of **5** petals   **3** groups of **10** petals
3 × **5** = 15   **3** × **10** = 30

Multiply.

A. 1 × 5 = _____   **F.** 6 × 5 = _____

B. 2 × 5 = _____   **G.** 7 × 5 = _____

C. 3 × 5 = _____   **H.** 8 × 5 = _____

D. 4 × 5 = _____   **I.** 9 × 5 = _____

E. 5 × 5 = _____   **J.** 10 × 5 = _____

K. 1 × 10 = _____   **P.** 6 × 10 = _____

L. 2 × 10 = _____   **Q.** 7 × 10 = _____

M. 3 × 10 = _____   **R.** 8 × 10 = _____

N. 4 × 10 = _____   **S.** 9 × 10 = _____

O. 5 × 10 = _____   **T.** 10 × 10 = _____

Look at your answers. What patterns do you see?

_____

Dive for Treasure

Number and Operations

Help the diver find the treasure. Write the answer to each problem on the path.

A. $2 \times 2 = $ _____

B. $3 \times 5 = $ _____

C. $2 \times 10 = $ _____

F. $4 \times 5 = $ _____

E. $5 \times 10 = $ _____

D. $3 \times 2 = $ _____

G. $1 \times 2 = $ _____

H. $3 \times 10 = $ _____

I. $5 \times 2 = $ _____

J. $7 \times 5 = $ _____

K. $8 \times 2 = $ _____

N. $7 \times 10 = $ _____

M. $6 \times 5 = $ _____

L. $4 \times 10 = $ _____

O. $9 \times 5 = $ _____

P. $8 \times 10 = $ _____

Q. $4 \times 2 = $ _____

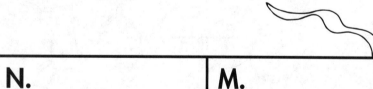

Bead Patterns

Algebra

Use the code to color the beads. Then complete each pattern.

Code			
r — red	b — blue	y — yellow	o — orange

A.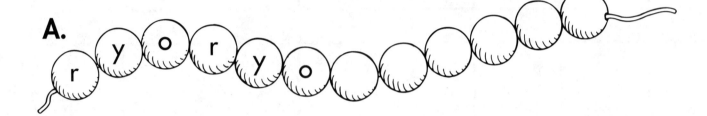

B.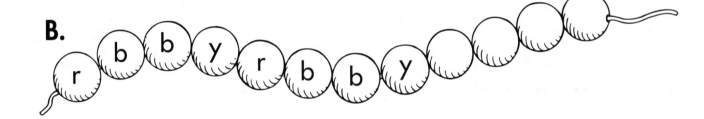

C.

D. Color the beads below to make your own colorful pattern.

Shape Patterns

Algebra

Draw shapes in the boxes to continue each pattern.

A.

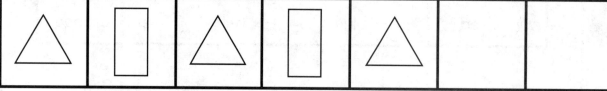

B.

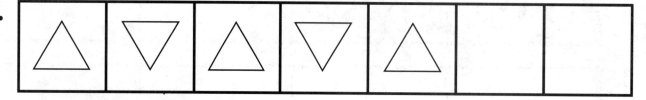

C.

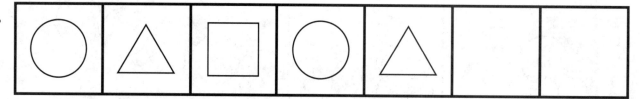

D.

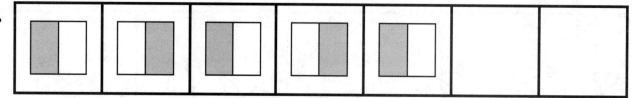

E.

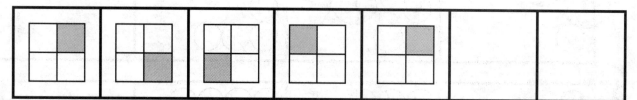

F.

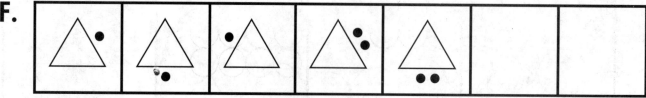

Standards-Based Math • 1–2 © 2004 Creative Teaching Press

Circle Add-Ons

Algebra

Draw circles to continue each pattern.

A.

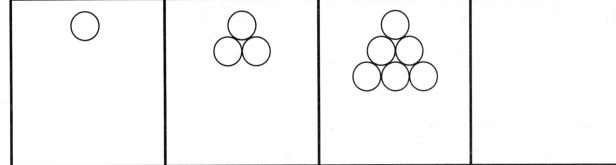

B.

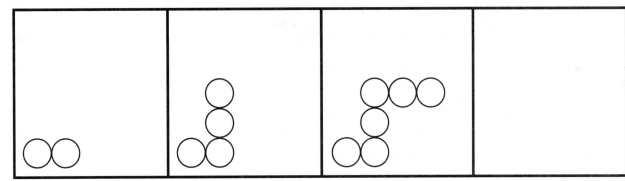

C.

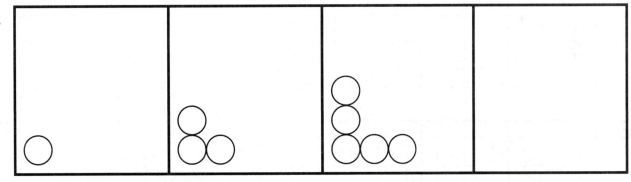

D.

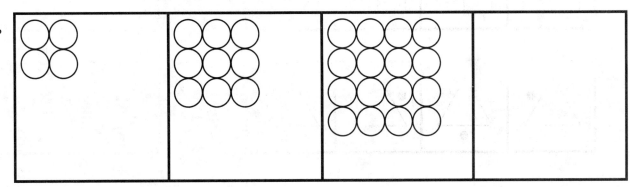

S-s-snaky Patterns

Algebra

Count by twos. Fill in the missing numbers.

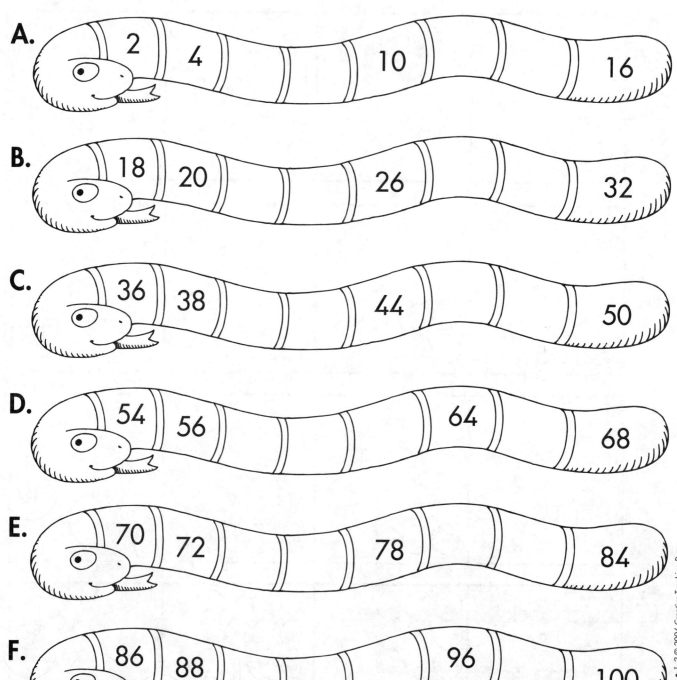

A. 2 4 ___ ___ 10 ___ ___ 16

B. 18 20 ___ ___ 26 ___ ___ 32

C. 36 38 ___ ___ 44 ___ ___ 50

D. 54 56 ___ ___ 64 ___ 68

E. 70 72 ___ ___ 78 ___ 84

F. 86 88 ___ ___ 96 ___ 100

Fishy Facts

Algebra

Look at the fish in each tank. Write two addition facts and two subtraction facts for each picture.

A.

$$4 + 2 = 6$$
$$2 + 4 = 6$$
$$6 - 2 = 4$$
$$6 - 4 = 2$$

B.

_____

_____

_____

_____

C.

_____

_____

_____

_____

D.

_____

_____

_____

_____

E.

_____

_____

_____

_____

F.

_____

_____

_____

_____

Name _____ Date _____

Domino Math

Algebra

Look at the dominoes and write the matching fact families.

A.

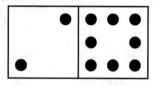

$$2 + 8 = 10$$
$$8 + 2 = 10$$
$$10 - 2 = 8$$
$$10 - 8 = 2$$

B.

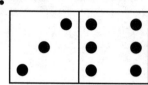

_____

_____

_____

_____

C.

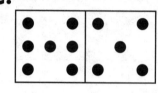

_____

_____

_____

_____

D.

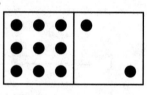

_____

_____

_____

_____

E.

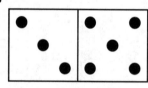

_____

_____

_____

_____

F.

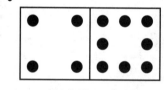

_____

_____

_____

_____

G.

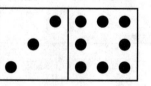

_____

_____

_____

_____

H.

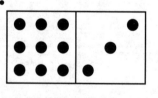

_____

_____

_____

_____

I.

_____

_____

_____

_____

Standards-Based Math • 1–2 © 2004 Creative Teaching Press

Number Houses

Algebra

Use the numbers on each house to write four related facts.

A. 6, 7, 13

$6 + 7 = 13$
$7 + 6 = 13$
$13 - 6 = 7$
$13 - 7 = 6$

B. 8, 7, 15

_____
_____
_____
_____

C. 9, 5, 14

_____
_____
_____
_____

D. 4, 9, 13

_____
_____
_____
_____

E. 8, 6, 14

_____
_____
_____
_____

F. 7, 9, 16

_____
_____
_____
_____

G. 6, 9, 15

_____
_____
_____
_____

H. 5, 8, 13

_____
_____
_____
_____

I. 9, 8, 17

_____
_____
_____
_____

Seesaw Sums

Algebra

Complete the facts so that both sides on each seesaw have the same sum.
Write the sum in the triangle below each seesaw.

A.

☐ + 1   3 + 2

B.

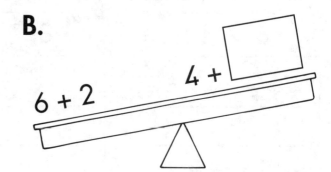

6 + 2   4 + ☐

C.

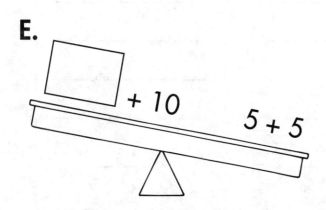

0 + 7   4 + ☐

D.

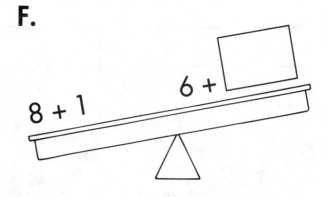

☐ + 2   3 + 1

E.

☐ + 10   5 + 5

F.

8 + 1   6 + ☐

G.

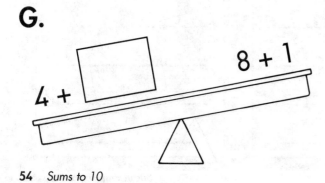

4 + ☐   8 + 1

H.

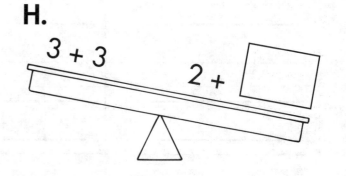

3 + 3   2 + ☐

Standards-Based Math • 1–2 © 2004 Creative Teaching Press

Seesaw Subtraction

Algebra

Complete the facts so that both sides on each seesaw have the same answer.
Write the answer in the triangle below each seesaw.

A.

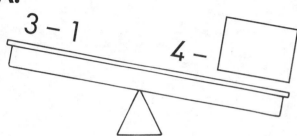

$3 - 1$

$4 -$

B.

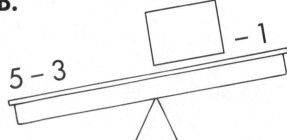

$- 1$

$5 - 3$

C.

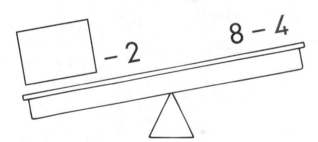

$- 2$

$8 - 4$

D.

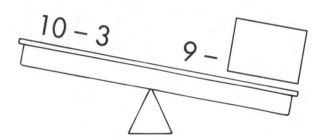

$10 - 3$

$9 -$

E.

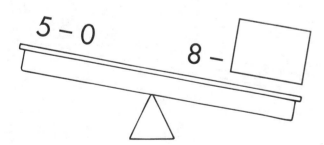

$5 - 0$

$8 -$

F.

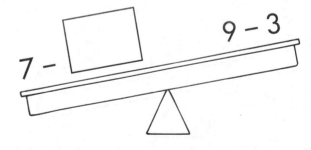

$9 - 3$

$7 -$

G.

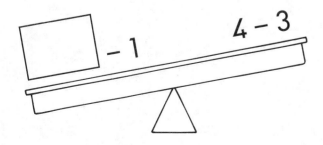

$- 1$

$4 - 3$

H.

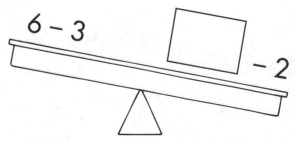

$6 - 3$

$- 2$

Seesaw Challenge

Algebra

Complete the facts so that both sides on each seesaw have the same answer.
Write the answer in the triangle below each seesaw.

A.

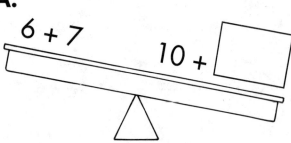

B.

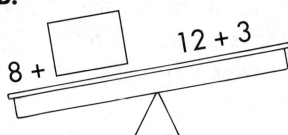

C.

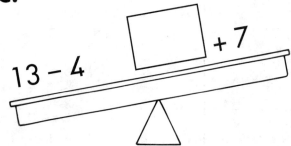

D.

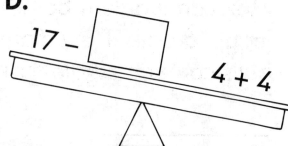

E.

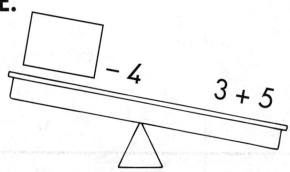

F.

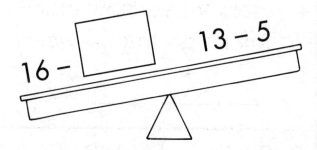

G.

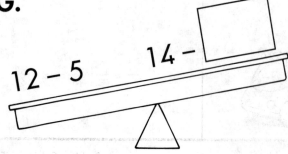

H.

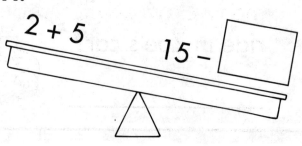

Standards-Based Math • 1–2 © 2004 Creative Teaching Press

Clowning Around

Algebra

Solve the problems. Draw a picture on the right to help you.

A. Max the Clown gave 10 peanuts to Fiesta the elephant. Fiesta ate 6 peanuts. How many peanuts were left?

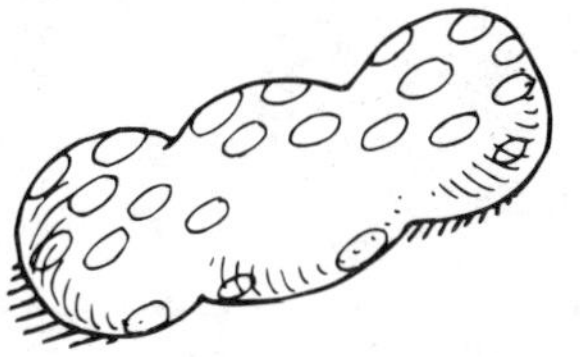

_____

B. Max can juggle 4 balls. Zoe can juggle 6 balls. How many more balls can Zoe juggle than Max?

_____

C. Max goes to 3 birthday parties every week. How many parties will he go to in 4 weeks?

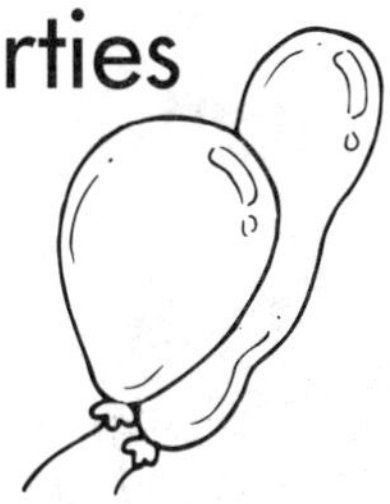

_____

D. Zoe's car holds 8 clowns. There are 15 clowns in the circus. How many clowns will not be able to ride in Zoe's car?

_____

Willie Wizard

Algebra

Solve the problems. Draw pictures on the right to help you.

A. Willie Wizard has 10 mice. He made 2 mice disappear. How many mice were left?

_____

B. Willie needs 8 lizards. He already has 3 lizards. How many more lizards does he need?

_____

C. Willie's cat eats 5 fish a day. How many fish will it eat in a week?

_____

D. Willie's frog can jump 2 feet at a time. How many times must it jump to go 10 feet?

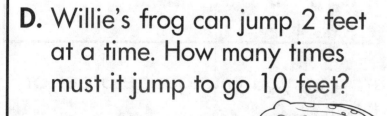

_____

Draw and Add

Algebra

Draw shapes to match the problems.
Write the answers to the problems.

A. 4 + 5 = _____ 	5 + 4 = _____
B. 3 + 7 = _____	7 + 3 = _____
C. 8 + 6 = _____	6 + 8 = _____
D. 7 + 5 = _____	5 + 7 = _____

What happens to the sum when you change the order of the numbers you are adding?

_____

Name _____   Date _____

Happy Birthday!
Algebra

Solve the problems. Draw pictures on the right to help you.

A. There are 12 children at a birthday party. There is the same number of boys and girls. How many boys are there? _____ How many girls are there? _____	
B. There are 12 gifts at the party. There are 4 more gifts with bows than there are gifts with no bows. How many gifts have bows? _____ How many gifts have no bows? _____	
C. There are 15 balloons. For every blue balloon, there are 2 red balloons. How many balloons are blue? _____ How many balloons are red? _____	
D. There are 10 candles on the birthday cake. Half of the candles are yellow. Two candles are orange. The rest are red. How many candles are yellow? _____ How many candles are red? _____	

Standards-Based Math • 1–2 © 2004 Creative Teaching Press

Name _____   Date _____

Piggy Bank Puzzlers

Algebra

Draw the coins to show how much money is in the piggy banks.

A. There are 3 coins. They add up to 20¢. 	**B.** There are 2 coins. They add up to 35¢. 
C. There are 3 coins. They add up to 45¢. 	**D.** There are 3 coins. They add up to 40¢. 
E. There are 4 coins. They add up to 45¢. 	**F.** There are 4 coins. They add up to 60¢. 
G. There are 4 coins. They add up to 80¢. 	**H.** There are 5 coins. They add up to 70¢. 

Missing Signs
Algebra

Write **+** or **−** in the circles to make the number sentences true.
Start at the left of each sentence. Then add or subtract in order.

A. 5 ◯ 1 ◯ 2 = 4

Make up five more problems
like the ones on this page.

B. 4 ◯ 4 ◯ 1 = 7

C. 3 ◯ 1 ◯ 3 = 5

D. 6 ◯ 2 ◯ 5 = 9

E. 8 ◯ 3 ◯ 4 = 7

F. 9 ◯ 4 ◯ 2 = 11

G. 12 ◯ 3 ◯ 3 = 6

H. 7 ◯ 6 ◯ 2 = 15

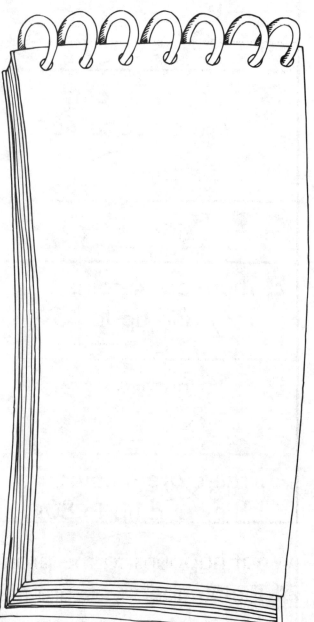

Standards-Based Math • 1–2 © 2004 Creative Teaching Press

Draw and Multiply

Algebra

Draw shapes to match the problems. Then solve the problems.

A.	2 groups of 4 △ △ △ △ △ △ △ △ 2 × 4 = _____	4 groups of 2 4 × 2 = _____
B.	3 groups of 5 3 × 5 = _____	5 groups of 3 5 × 3 = _____
C.	4 groups of 3 4 × 3 = _____	3 groups of 4 3 × 4 = _____
D.	5 groups of 2 5 × 2 = _____	2 groups of 5 2 × 5 = _____

What happens to the product when you change the order of the numbers you are multiplying?

_____

Hidden Facts

Algebra

Circle the hidden multiplication facts.

Add an ✕ and an = for each fact.

A. 3 $\boxed{5 \times 2 = 10}$ 8   **K.** 2   3   5   15   6

B. 2   3   1   5   5   **L.** 6   2   12   15   20

C. 4   8   2   2   4   **M.** 7   2   10   7   70

D. 5   4   10   40   10   **N.** 1   2   6   5   30

E. 2   5   10   30   5   **O.** 3   5   5   25   15

F. 1   2   2   6   6   **P.** 4   9   5   45   20

G. 4   2   10   5   50   **Q.** 4   3   2   6   10

H. 8   5   40   10   4   **R.** 1   5   8   2   16

I. 2   4   5   20   10   **S.** 9   2   18   2   14

J. 6   2   8   10   80   **T.** 5   6   10   60   30

Standards-Based Math • 1–2 © 2004 Creative Teaching Press

Bagfuls of Numbers

Algebra

Look at the numbers in each bag. Use them to write the numbers that match the clues. Do not use a number in a bag more than once.

A. The number is odd.
It is greater than 40.
It is less than 45.

_____

B. The number is even.
It is greater than 55.
It is less than 75.

_____

C. The number is even.
It is greater than 35.
It is less than 50.

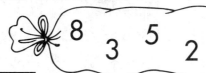

_____

D. The number is odd.
It is greater than 45.
It is less than 60.

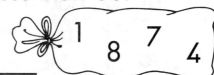

_____

E. The number is even.
It is greater than 60.
It is less than 70.

_____

F. The number is odd.
It is greater than 25.
It is less than 30.

_____

G. The number is odd.
It is greater than 60.
It is less than 65.

_____

H. The number is even.
It is greater than 75.
It is less than 85.

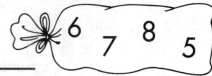

_____

Name _____ Date _____

Evan's Riddles

Algebra

Evan wrote some riddles about two-digit numbers.
Read his clues. Then write the number that answers each riddle.

A. Both of my digits are even. Add my digits and you get 6. I am greater than 30.

What number am I?

_____

B. Both of my digits are odd. The ones digit is 8 more than my tens digit.

What number am I?

_____

C. I am an odd number. The sum of my digits is 9. I am less than 30.

What number am I?

_____

D. My tens digit is 2 more than my ones digit. Add my digits and you get 10.

What number am I?

_____

Write your own riddle about a two-digit number. Ask a friend to solve it.

Standards-Based Math • 1–2 © 2004 Creative Teaching Press

Sides and Corners

Geometry

Look at each shape. Count the number of sides and corners.

A.
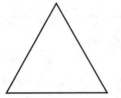

_____ sides _____ corners

B.

_____ sides _____ corners

C.
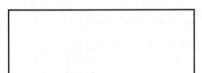

_____ sides _____ corners

D.
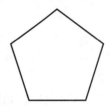

_____ sides _____ corners

E.

_____ sides _____ corners

F.
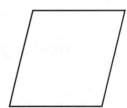

_____ sides _____ corners

G.
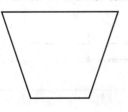

_____ sides _____ corners

H.
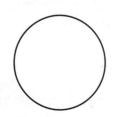

_____ sides _____ corners

Name _____ Date _____

A Shape Picture

Geometry

Use the code to color the picture.

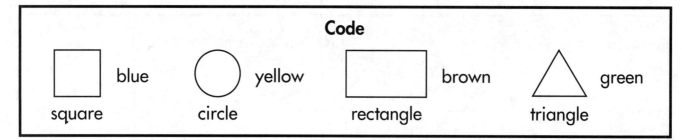

Code

□ blue	○ yellow	▭ brown	△ green
square	circle	rectangle	triangle

Write how many shapes you found in the picture.

□ _____ ○ _____ ▭ _____ △ _____

Dot-to-Dot Shapes

Geometry

Use a ruler to connect the dots in order.
Look at the shape you made.
Write its name on the line.
Use the words in the Word Box to help you.

Word Box		
square	rectangle	hexagon
triangle	pentagon	octagon

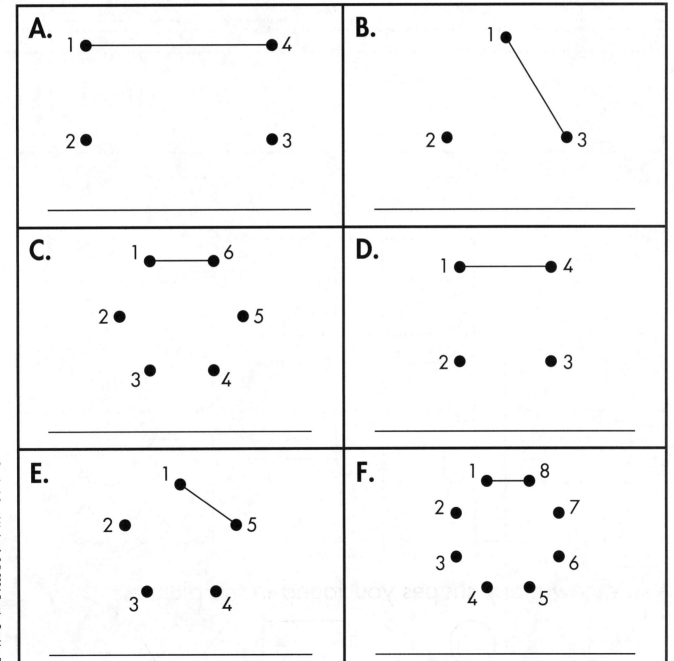

Same Size and Shape

Geometry

Shapes that are the same size and shape are **congruent**.
Triangle A and Triangle B are congruent.
Triangle C is not congruent. It is the same shape,
but it is not the same size.

A   B   C

Look at the first shape in each row. Circle the shape that is congruent to it.

A.

B.

C.

D.

E.

Name _____ Date _____

Sets of Shapes

Geometry

Sam and Pam sorted eight paper shapes into two different sets.

Sam's Sets

Set 1

Set 2

How did Sam decide if a shape fit in Set 1 or 2?

_____

_____

Pam's Sets

Set 1

Set 2

How did Pam decide if a shape fit in Set 1 or 2?

_____

_____

Now it's your turn!

Sort the shapes in a different way. Draw your sets in the box. Tell how you decided if a shape belonged in Set 1 or Set 2.

_____

_____

Set 1

Set 2

Standards-Based Math • 1–2 © 2004 Creative Teaching Press

Pairs of Shapes

Geometry

A. Color the two shapes that can be put together to form a rectangle.

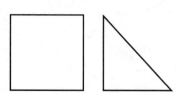 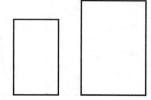 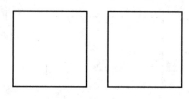

B. Color the two shapes that can be put together to form a triangle.

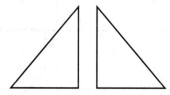 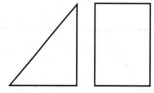 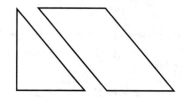

C. Color the two shapes that can be put together to form a square.

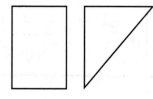 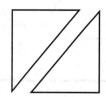 

D. Color the two shapes that can be put together to form a trapezoid.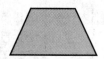

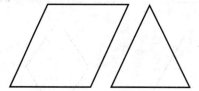 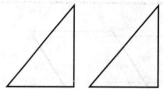 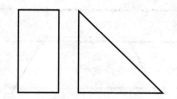

Name ______________________________ Date ____________

Put Them Together

Geometry

Look at the first shape in each row. Circle the two shapes that can be joined together to make the first shape.

A.

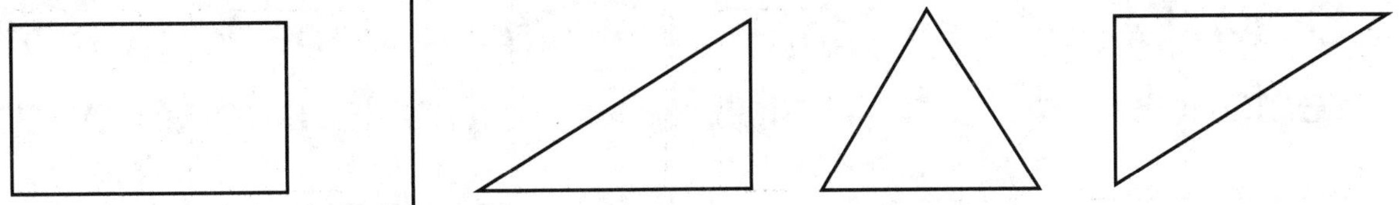

B.

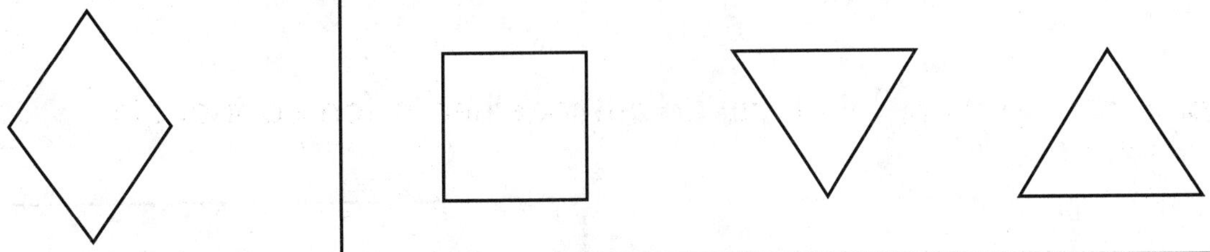

C.

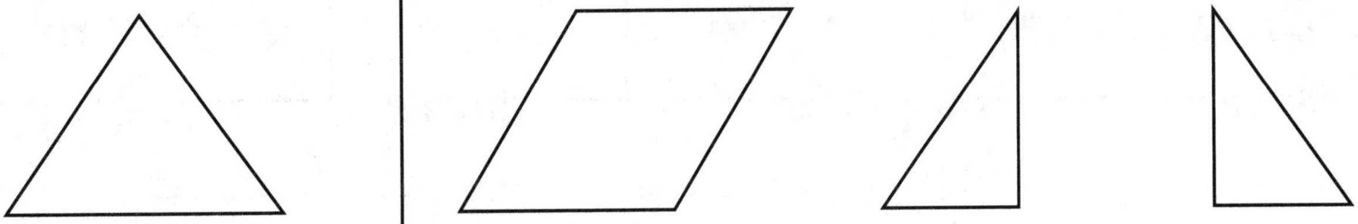

D.

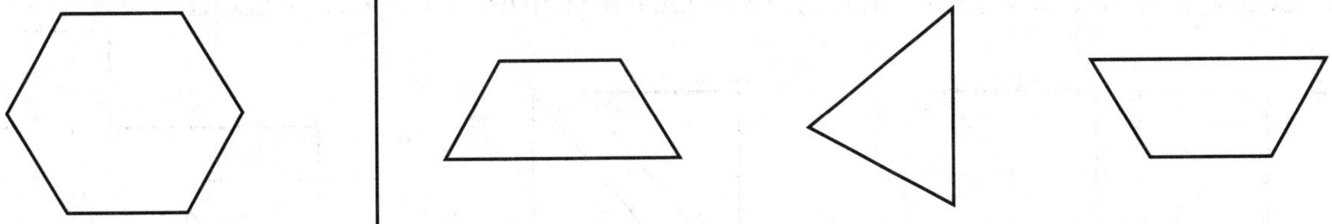

E.

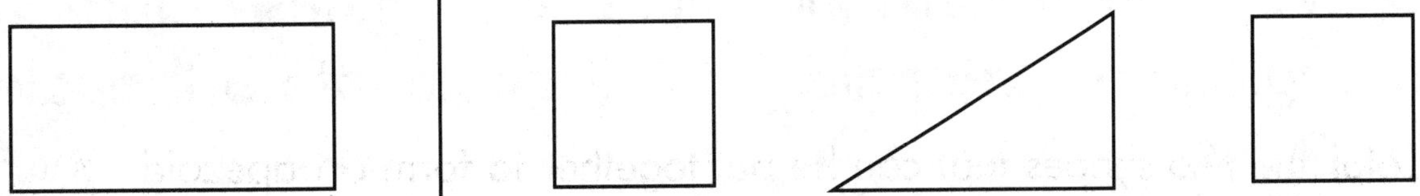

F.

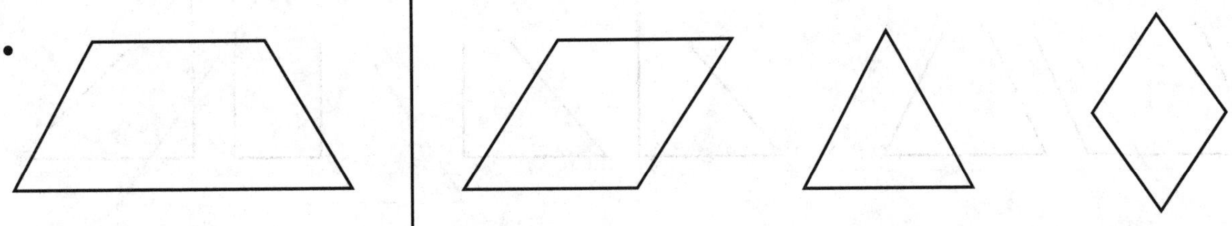

Abracadabra!

Geometry

Help Marvo the Magician make new shapes.
Use a pencil and a ruler to draw lines on the shapes below.

A. Draw 1 line to turn the rectangle into 2 triangles.

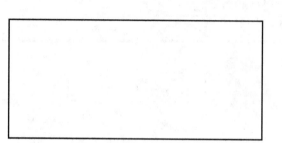

B. Draw 1 line to turn the rectangle into 2 squares.

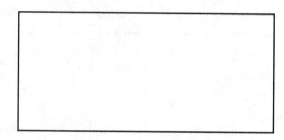

C. Draw 2 lines to turn the square into 4 squares.

D. Draw 2 lines to turn the square into 4 triangles.

E. Draw 2 lines to turn the triangle into 3 triangles.

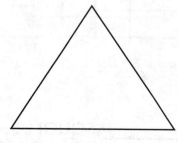

F. Draw 2 lines to turn the triangle into 3 triangles.

(Make your drawing different from E.)

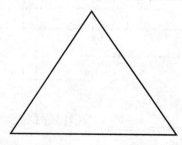

Tricky Squares

Geometry

Count the total number of squares you see in each design.
Be careful. Some squares may be hidden!

A.

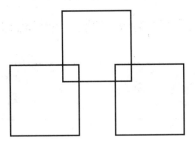

_____ squares

B.

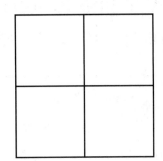

_____ squares

C.

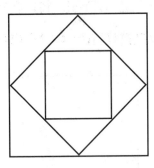

_____ squares

D.

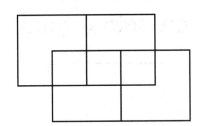

_____ squares

E.

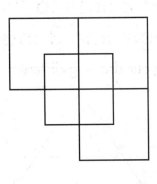

_____ squares

F.

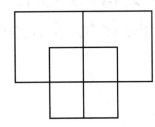

_____ squares

Tricky Triangles

Geometry

There are three triangles in this design. There are two small triangles and one large triangle.

Count the total number of triangles you see in each design. Be careful. Some triangles may be hidden!

A.

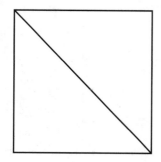

_____ triangles

B.

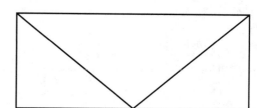

_____ triangles

C.

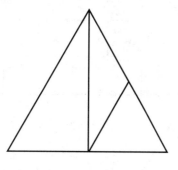

_____ triangles

D.

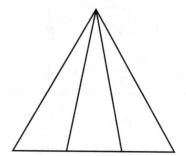

_____ triangles

E.

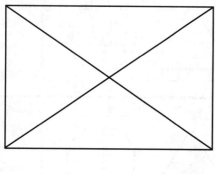

_____ triangles

F.

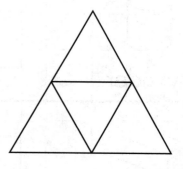

_____ triangles

Standards-Based Math • 1–2 © 2004 Creative Teaching Press

Name _____   Date _____

Faces and Shapes

Geometry

Look at the shaded face of each solid.
Color the shape that matches the face.

A. cube	▭ △ ○ □
B. pyramid	▭ △ ○ □
C. rectangular prism	▭ △ ○ □
D. pyramid	▭ △ ○ □
E. cylinder	▭ △ ○ □
F. rectangular prism	▭ △ ○ □

Blocks and Bases

Geometry

Suppose you picked up each of the toy blocks below.
What would its base look like? (The base is the bottom part of
the block.) Write the name of the shape that matches the base.
(You may choose a shape more than once.)

 square   circle   rectangle

A.  base shape _____	**B.** 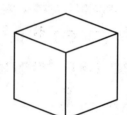 base shape _____
C. 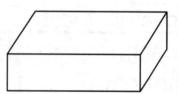 base shape _____	**D.**  base shape _____
E. 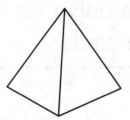 base shape _____	**F.** 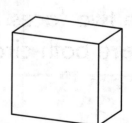 base shape _____

What Am I?

Geometry

Read the clues. Then write the name of the matching solid.

  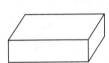

cube   cone   rectangular prism   sphere   cylinder   pyramid

A. I have faces that are shaped like triangles. What am I?

_____

B. I have six faces. Each face is a square. What am I?

_____

C. I have the same number of sides and corners as a cube. What am I?

_____

D. I have no faces. I only have curved sides. What am I?

_____

E. I have two faces. They are both circles. What am I?

_____

F. I have only one face. It is a circle. What am I?

_____

Exact Halves

Geometry

When a shape has symmetry, it can be divided in half so that the halves match exactly. The shape is said to be **symmetrical**. The line that divides the shape in half is called the **line of symmetry**.

This shape has symmetry.

Draw a line to divide each shape into two equal parts.

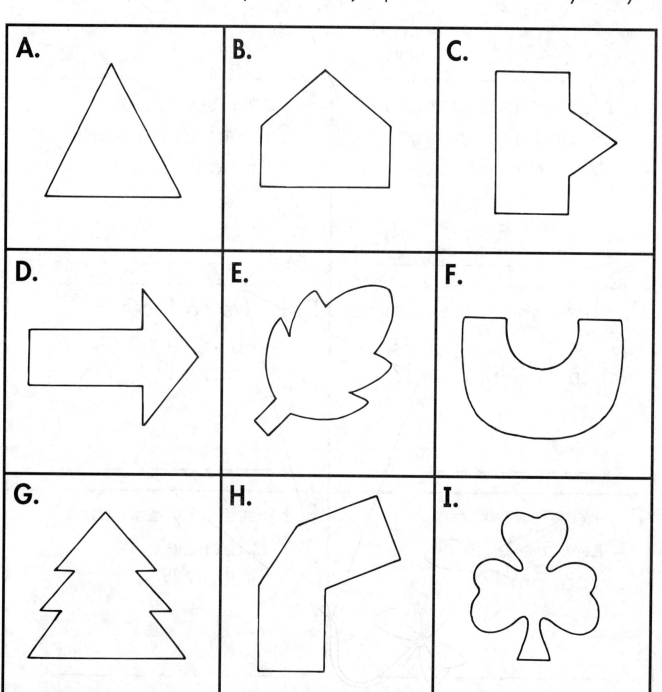

Kite Designs

Geometry

Complete each kite so that the halves match. Color your kite designs.

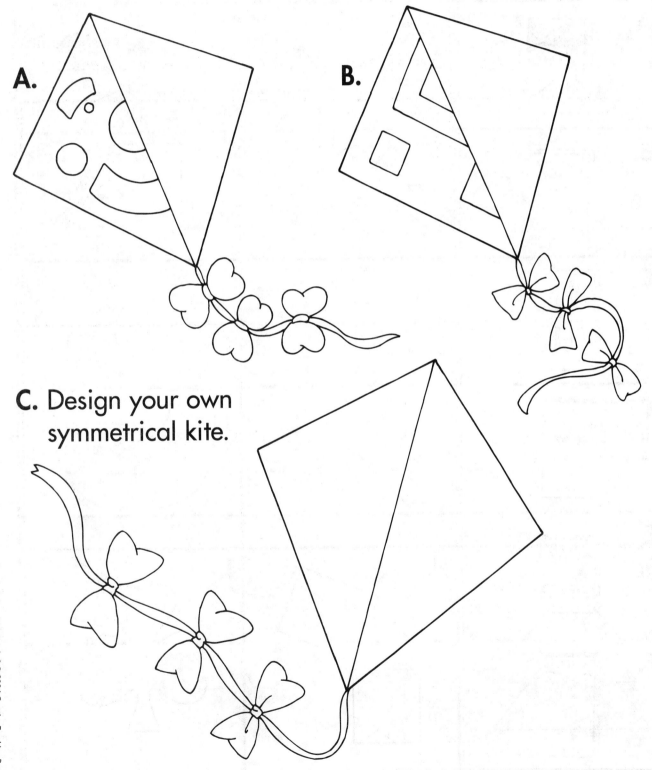

A.

B.

C. Design your own symmetrical kite.

Name _____   Date _____

Paper Cutouts

Geometry

Look at each row. Circle the shape that matches the paper it was cut from.

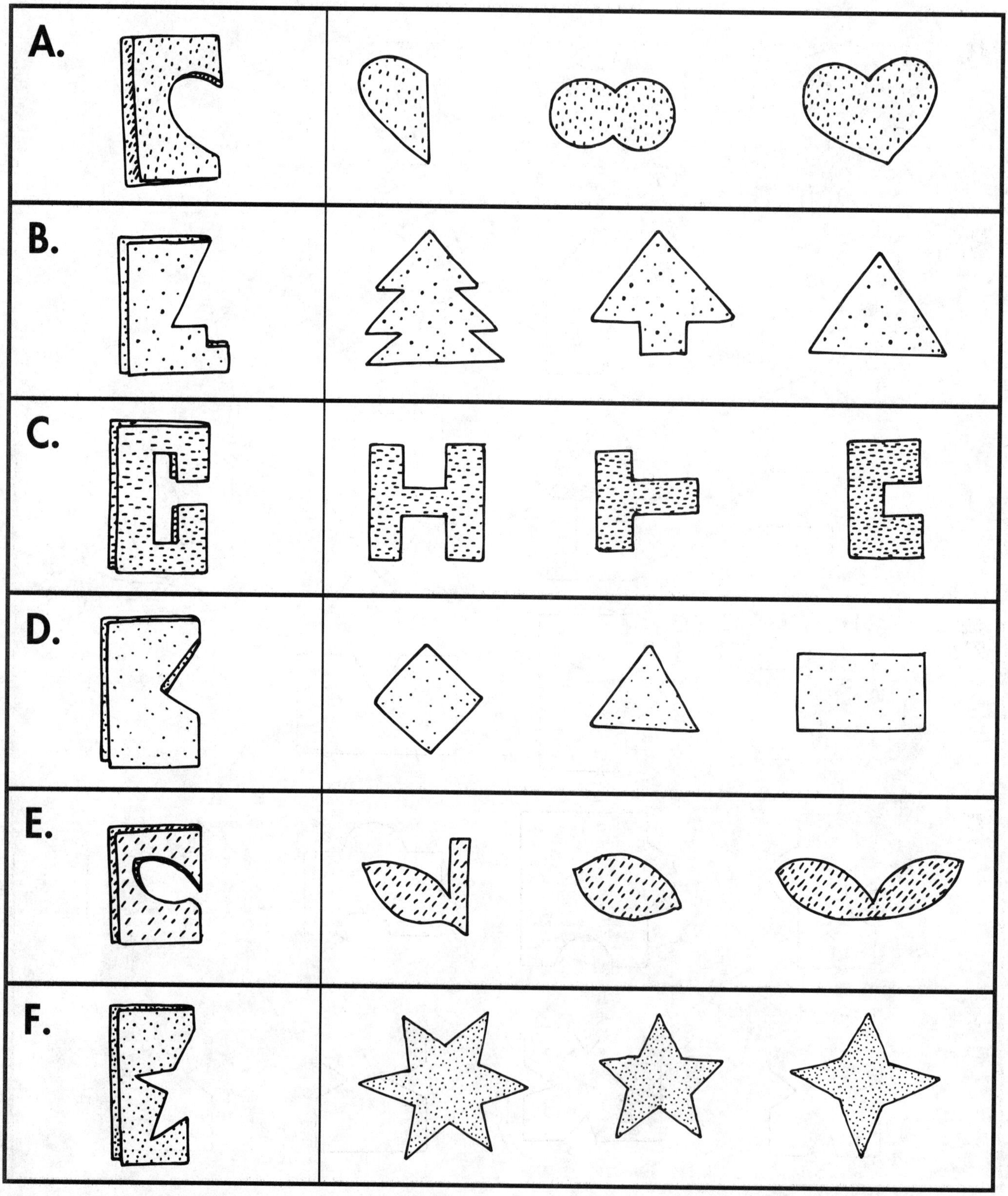

Standards-Based Math • 1–2 © 2004 Creative Teaching Press

Name ______________________________ Date ____________

Which Shape Doesn't Belong?

Geometry

Cross out the shape in each row that does not belong.

A.

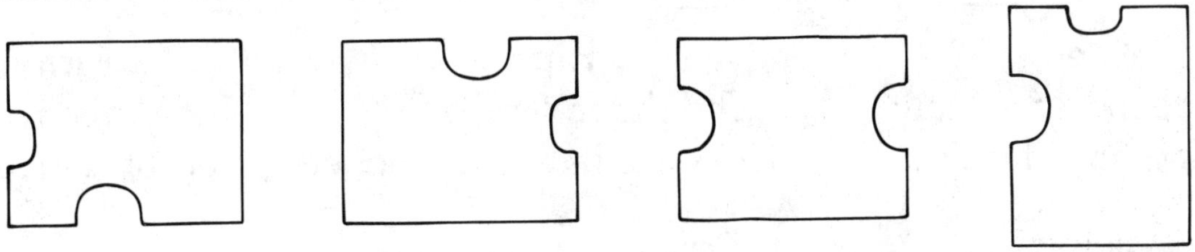

B.

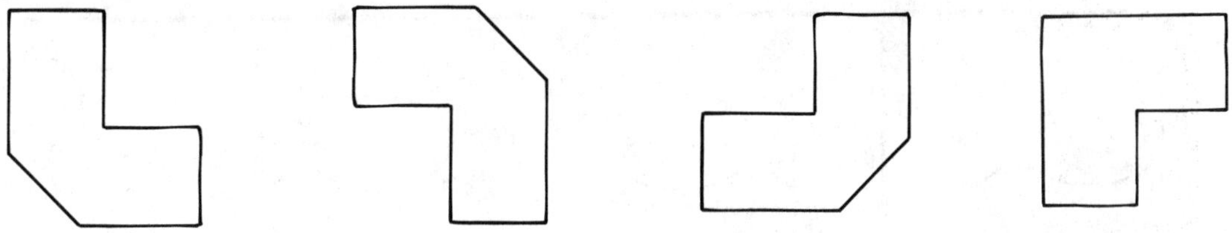

C.

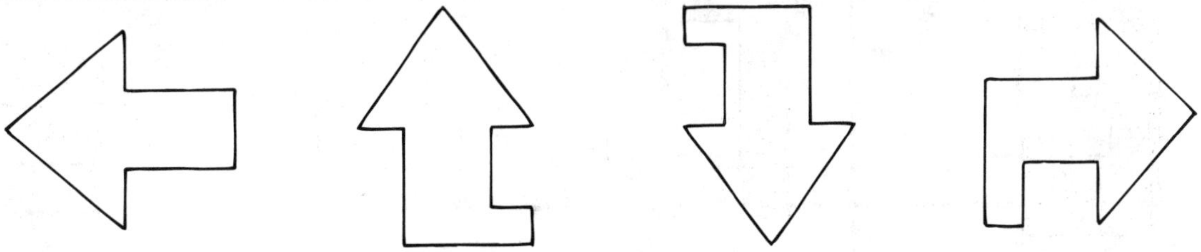

D.

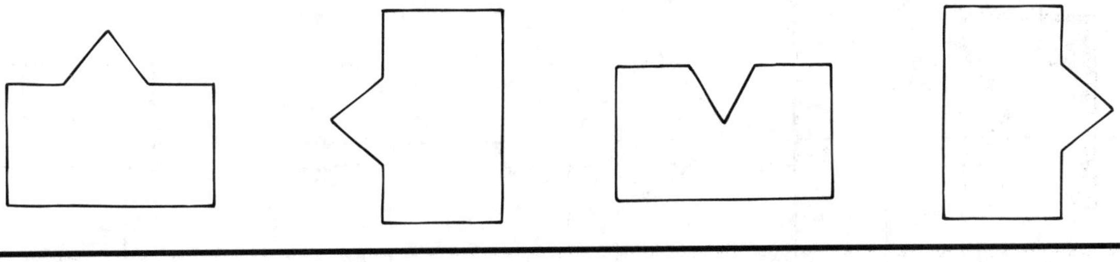

E.

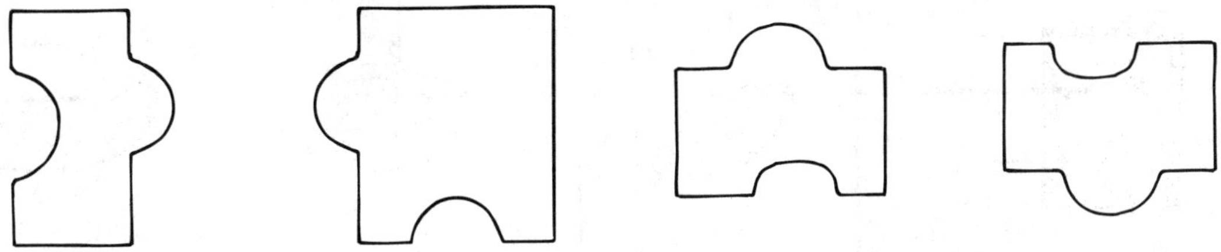

F.

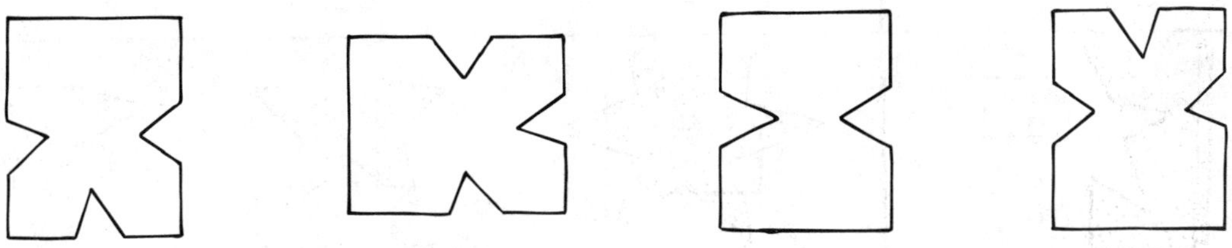

Slides, Flips, and Turns

Geometry

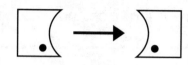

Slide
The shape moves without changing how it looks.

Flip
The shape flips over. It looks like its mirror image.

Turn
The shape rotates, or turns around.

Look at the first shape in each row. Circle the slide, flip, or turn.

Slide **A.**			
B.			
Flip **C.**			
D.			
Turn **E.**			
F.			

Find the Shape

Geometry

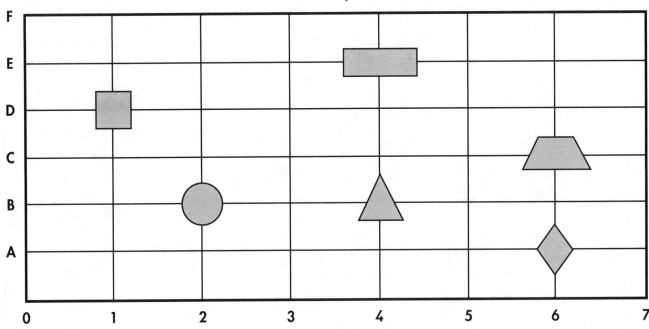

Draw the shape for each location described below. To find the shape, first put your finger on the line that has the correct letter. Then move your finger across that line until you come to the line with the correct number.

A. What shape is found at B2?

_____

B. What shape is found at E4?

_____

C. What shape is found at A6?

_____

D. What shape is found at C6?

_____

E. What shape is found at D1?

_____

F. What shape is found at B4?

_____

Name _____ Date _____

Where Is the Toy?

Geometry

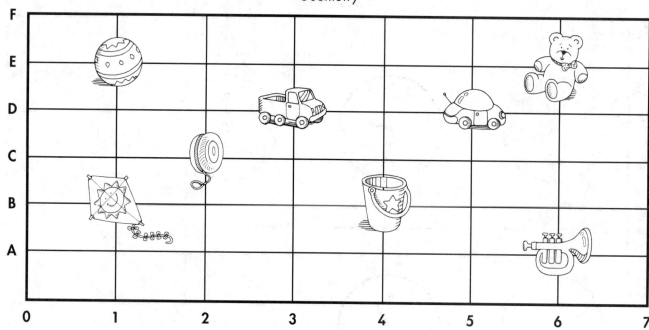

Write where each toy can be found. Write the letter and line that tell which two lines the toy is on. (Example: A2)

A. Where is the ?

_____

B. Where is the  ?

_____

C. Where is the ?

_____

D. Where is the 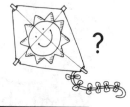 ?

_____

E. Where is the  ?

_____

F. Where is the ?

_____

G. Where is the  ?

_____

H. Where is the ?

_____

What Time Is It?

Measurement

Write the time.

A.

_____ _____ _____ _____

B.

_____ _____ _____ _____

C.

_____ _____ _____ _____

D.

_____ _____ _____ _____

Feeding Time

Measurement

Zookeeper Zed feeds the animals at the zoo.
Write the time he does each chore.

A. Peel bananas.

_____

B. Feed the monkeys.

_____

C. Catch fish.

_____

D. Feed the bears.

_____

E. Feed the seals.

_____

F. Chop some hay.

_____

G. Feed the elephants.

_____

H. Pick some leaves.

_____

I. Feed the koalas.

_____

When do you eat lunch? Draw
hands on the clock to show your
answer. Then write the time.

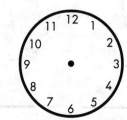

_____

Standards-Based Math • 1–2 © 2004 Creative Teaching Press

Watch the Clock

Measurement

1:45
quarter to 2

Write the time
two ways.

2:15
quarter past 2

A.

_____

_____

B.

_____

_____

C.

_____

_____

D.

_____

_____

E.

_____

_____

F.

_____

_____

G.

_____

_____

H.

_____

_____

I.

_____

_____

Name _____   Date _____

Passing Time

Measurement

Solve each problem. Write the time.
Then draw hands on the clock to match.

A. The movie started at 3:00. It lasted 2 hours. What time did the movie end? _____ 	**B.** Meg went to Kim's house at 11:30. She stayed for 3 hours. What time did Meg leave? _____ 
C. Lynn sleeps for 9 hours each night. She goes to bed at 9:00. What time does she get up? _____ 	**D.** Matt read for half an hour. He finished at 7:45. What time did Matt start reading? _____ 
E. Mr. Mays left home at 10:30. He returned 5 hours later. What time did he get home? _____ 	**F.** Mrs. Lum shopped for half an hour. She finished at 2:15. What time did she start shopping? _____

Standards-Based Math • 1–2 © 2004 Creative Teaching Press

Here Come the Months

Measurement

There are 12 months in the year.
Use the chart at the right to help you solve the problems.

A. Mandy's birthday is in June. Kelly's birthday is three months later. In what month is Kelly's birthday?

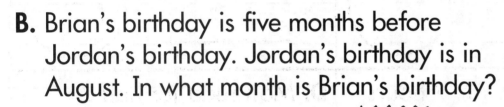

_____

B. Brian's birthday is five months before Jordan's birthday. Jordan's birthday is in August. In what month is Brian's birthday?

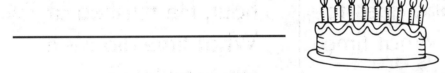

_____

C. Mrs. Garcia's son turned eight months old in October. In what month was he born?

_____

D. Joel visited his uncle for two months. He returned home at the beginning of September. In what month did he arrive at his uncle's house?

_____

Months of the Year
January
February
March
April
May
June
July
August
September
October
November
December

May Days

Measurement

Use the calendar to solve the problems.

May

Sunday	Monday	Tuesday	Wednesday	Thursday	Friday	Saturday
			1	2	3	4
5	6	7	8	9	10	11
12	13	14	15	16	17	18
19	20	21	22	23	24	25
26	27	28	29	30	31	

A. On what day is the first day in May? _____

B. How many days are in May? _____

C. What day is May 7? _____

D. What day is May 26? _____

E. How many Mondays are in the month? _____

F. What is the date of the
third Saturday in the month? _____

G. What is the date of the
last Wednesday in the month? _____

H. Which days of the week occur
five times in the month? _____

Stringing Along

Measurement

How long is each string? First write your guess on the chart.
Then use an inch ruler to measure each string.

1 inch

A.

B.

	My Guess (in inches)	**My Check** (in inches)
A.		
B.		
C.		
D.		
E.		
F.		

C.

D.

E.

F.

Going to the Fair

Measurement

Use an inch ruler to measure the parts of the path.

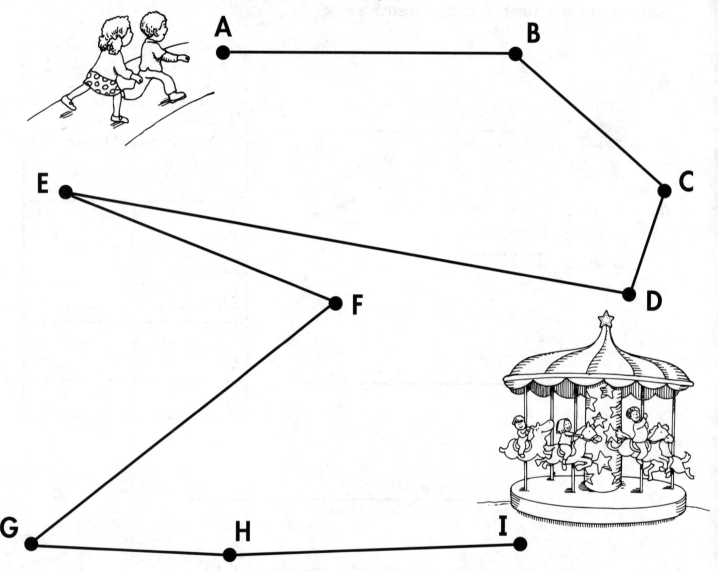

A. From A to B _____ inches   **E.** From E to F _____ inches

B. From B to C _____ inches   **F.** From F to G _____ inches

C. From C to D _____ inches   **G.** From G to H _____ inches

D. From D to E _____ inches   **H.** From H to I _____ inches

Standards-Based Math • 1–2 © 2004 Creative Teaching Press

Centimeter Caterpillars

Measurement

How long is each caterpillar? First write your guess on the chart. Then use a centimeter ruler to measure each string.

— 1 centimeter

A.

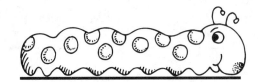

B.

C.

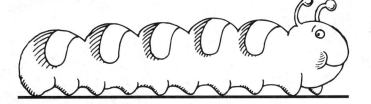

D.

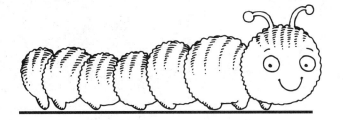

	My Guess (in centimeters)	My Check (in centimeters)
A.		
B.		
C.		
D.		
E.		
F.		

E.

F.

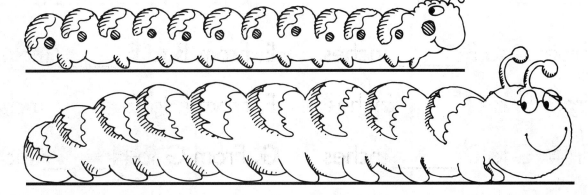

Ounces and Pounds

Measurement

Write **ounces** or **pounds** to show how much each object weighs.

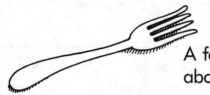 A fork weighs about 1 ounce.

 A shoe weighs about 1 pound.

A.

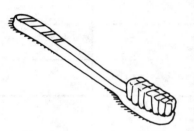

about 2 _____

B.

about 6 _____

C.

about 10 _____

D.

about 2 _____

E.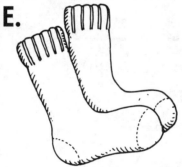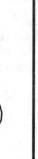

about 4 _____

F.

about 2 _____

G.

about 7 _____

H.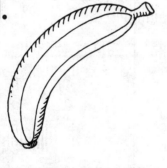

about 8 _____

I.

about 35 _____

Standards-Based Math • 1–2 © 2004 Creative Teaching Press

Grams and Kilograms

Measurement

A gram is used to measure the weight of light objects.
A kilogram is used to measure the weight of heavier objects.

What would best describe the weight of the objects below?
Write **grams** or **kilograms**.

A paper clip weighs
about 1 gram.

A tape holder weighs
about 1 kilogram.

A.	B.	C.
_____	_____	_____
D.	E.	F.
_____	_____	_____
G.	H.	I.
_____	_____	_____

Name _____ Date _____

One pint equals two cups.

Cups and Pints

Measurement

Look at the pictures. How many cups can be filled?
Color the cups and fill in the blanks to show your answer.

A.

1 pint = _____ cups

B.

2 pints = _____ cups

C.

3 pints = _____ cups

D.

4 pints = _____ cups

Standards-Based Math • 1–2 © 2004 Creative Teaching Press

Cups, Pints, and Quarts

Measurement

2 cups = 1 pint   2 pints = 1 quart   4 cups = 1 quart

Which holds more? Circle your answer.

A. 2 cups or 2 pints?

B. 1 quart or 5 cups?

C. 2 quarts or 3 pints?

D. 3 pints or 4 cups?

E. 3 pints or 1 quart?

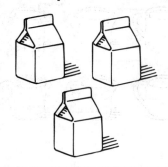

F. 4 cups or 2 quarts?

Equal Measures

Measurement

 =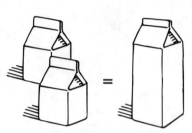

2 cups = 1 pint   2 pints = 1 quart   4 quarts = 1 gallon

Fill in the blanks.

A.

1 pint = _____ cups

2 pints = _____ cups

3 pints = _____ cups

4 pints = _____ cups

B.

1 quart = _____ pints

1 quart = _____ cups

2 quarts = _____ pints

2 quarts = _____ cups

C.

1 gallon = _____ quarts

2 gallons = _____ quarts

3 gallons = _____ quarts

4 gallons = _____ quarts

D.

1 gallon = _____ pints

1 gallon = _____ cups

2 gallons = _____ pints

2 gallons = _____ cups

Liquid Measures

Measurement

Answer each problem with the correct unit of measure.

A. Sara is having some soup. Will she drink about 1 cup or 1 quart of soup?

_____

B. Kent is making some pudding. He needs 3 cups of milk. Should he buy 1 pint or 1 quart of milk?

2 cups = 1 pint
2 pints = 1 quart
4 quarts = 1 gallon

_____

C. Lauren has invited 8 friends to her home. She will make some juice for them. Will she need 1 quart or 1 gallon of juice?

_____

D. Evan poured some lemonade for himself and a friend. Did he pour about 1 pint or 1 quart of lemonade in all?

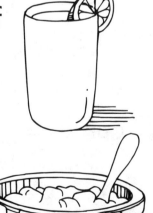

_____

E. Mrs. Lee made a pot of stew for her family. Did she use about 1 pint or 1 gallon of water for the stew?

_____

Name _____ Date _____

Square Inches

Measurement

square inch

Area is the measure of a surface. It is the number of square units that are needed to cover the surface of each shape. One type of square unit is the square inch.

Find the area of each shape. First connect the dots to make one-inch squares. Then count the squares and write the area.

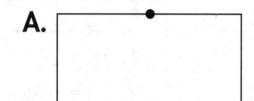

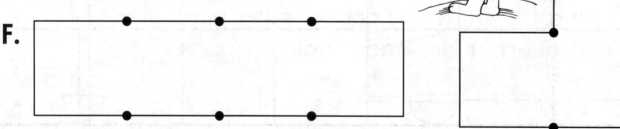

A. _____ square inches   B. _____ square inches

C. _____ square inches   D. _____ square inches

E. _____ square inches   F. _____ square inches

Standards-Based Math • 1–2 © 2004 Creative Teaching Press

Find the Area

Measurement

Find the area for each shape. Count the shaded squares. Do not count a square if less than half of it is shaded.

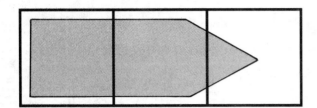

The area of this shape is 2 square inches.

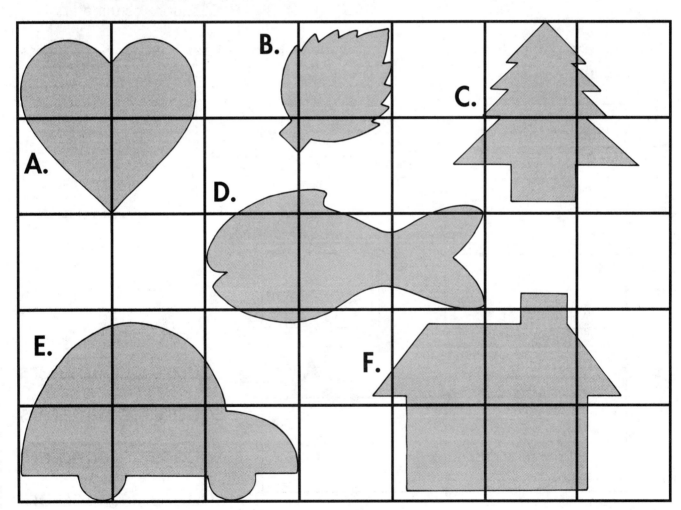

A. _____ square inches   **B.** _____ square inches

C. _____ square inches   **D.** _____ square inches

E. _____ square inches   **F.** _____ square inches

Square Centimeters

Measurement

The area of a surface of a shape can be measured in square centimeters.

square
centimeter

The area of this shape is
6 square centimeters.

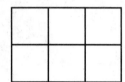

 The area of this shape is
2 square centimeters.

Find the area of each shape. First connect the dots to make one-centimeter squares.
Then count the squares and write the area.

A.

B.

C.

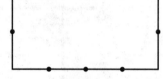 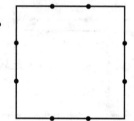 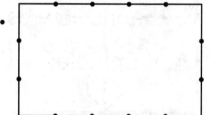

D.

E.

 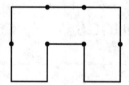

F.

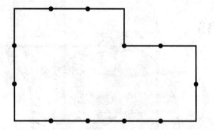

A. ____ square centimeters

B. ____ square centimeters

C. ____ square centimeters

D. ____ square centimeters

E. ____ square centimeters

F. ____ square centimeters

G.

G. ____ square centimeters

Standards-Based Math • 1–2 © 2004 Creative Teaching Press

Units of Measure

Measurement

Complete each sentence with the word that makes the most sense. Write it on the line.

A. My book is 8 _____ wide.

    inches   feet   yards

B. Cory's cat weighs 35 _____.

    cups   pounds   pints

C. The movie was 2 _____ long.

    minutes   hours   days

D. This banana weighs about 8 _____.

    ounces   pounds   cups

E. Mom poured 2 _____ of water into the teapot.

    pounds   cups   gallons

F. Mr. Henson is 6 _____ tall.

    inches   feet   yards

G. I have to bake the cookies for 15 _____.

    minutes   hours   days

Sky Gazing

Data Analysis and Probability

Help Willie the Wizard count the objects in the sky. Draw tally marks for each object.

Write how many objects you counted.

_____   _____   _____   _____

Standards-Based Math • 1–2 © 2004 Creative Teaching Press

Name _____ Date _____

Show Time

Data Analysis and Probability

Coco the Clown needs to gather her things. Draw tally marks for each object.

Use your tally marks to help you answer the questions.

A. How many balls are there? _____

B. Are there more balloons or bows? _____

C. How many more bows are there than hats? _____

Standards-Based Math • 1–2 © 2004 Creative Teaching Press

Name _____ Date _____

All Kinds of Weather

Data Analysis and Probability

Chelsea made a bar graph to keep track of the weather.

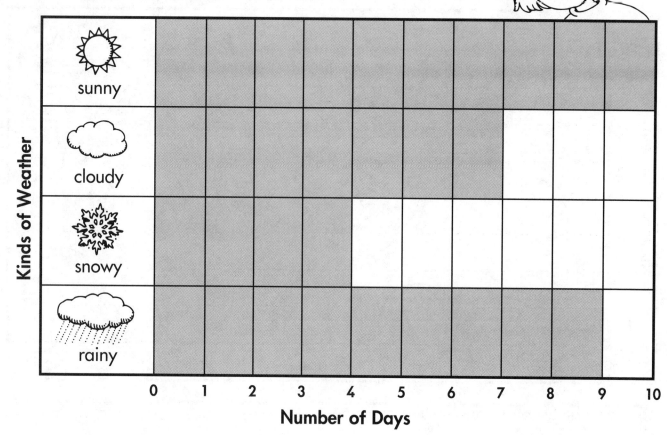

Use the bar graph to answer the questions.

A. How many days were sunny? _____

B. How many days were snowy? _____

C. How many more sunny days
were there than cloudy days? _____

D. How many days were
either snowy or rainy? _____

E. How many more rainy days
were there than snowy days? _____

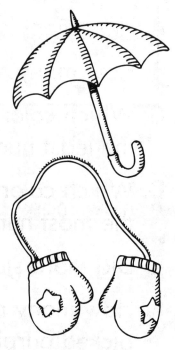

Standards-Based Math • 1–2 © 2004 Creative Teaching Press

Name _____ Date _____

A Color Graph
Data Analysis and Probability

Mrs. Mann's students made a graph to show what colors they liked the best.

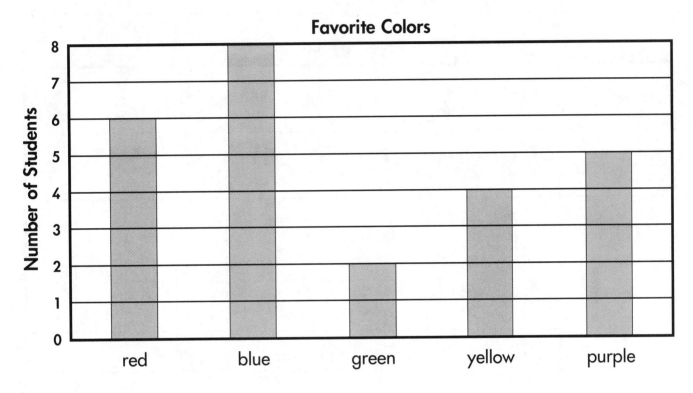

Favorite Colors

Use the bar graph to answer the questions.

A. How many students liked purple best? _____

B. How many students liked red best? _____

C. Which color was picked by
the least number of students? _____

D. Which color was picked by
the most number of students? _____

E. Did more students pick red or yellow? _____

F. How many more students
picked purple than green? _____

A Fruity Graph

Data Analysis and Probability

Miss Reed's students voted on which fruits they liked the best.
The picture graph shows how they voted.

Favorite Fruits

apple	😊 😊 😊 😊 😊 😊 😊 😊
orange	😊 😊 😊 😊 😊
grapes	😊 😊
pear	😊 😊 😊
banana	😊 😊 😊 😊 😊 😊

😊 = 1 student

A. How many students liked oranges the best? _____

B. How many students liked grapes the best? _____

C. Which fruit got the most votes? _____

D. Did more students like pears or bananas? _____

E. How many students voted
for either apples or oranges? _____

F. How many students voted in all? _____

Standards-Based Math • 1–2 © 2004 Creative Teaching Press

Name _____ Date _____

Zoo Animals

Data Analysis and Probability

Cindy went to the zoo. She kept track of the animals she saw.
The graph shows some of the animals she saw.

Zoo Animals

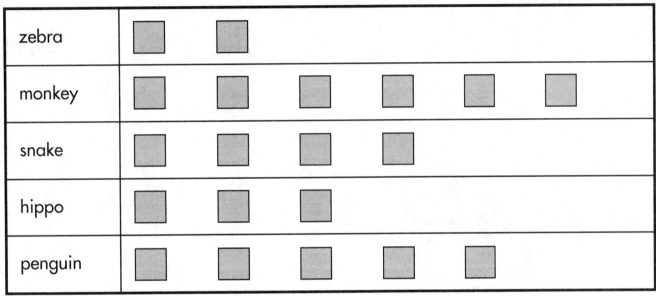

zebra	▨ ▨
monkey	▨ ▨ ▨ ▨ ▨ ▨
snake	▨ ▨ ▨ ▨
hippo	▨ ▨ ▨
penguin	▨ ▨ ▨ ▨ ▨

▨ = 2 animals

A. How many zebras did Cindy see? _____

B. How many monkeys did Cindy see? _____

C. Were there more penguins or snakes? _____

D. How many hippos and
zebras were there altogether? _____

E. How many more monkeys
were there than snakes? _____

F. How many more hippos
were there than zebras? _____

Standards-Based Math • 1–2 © 2004 Creative Teaching Press

Reading a Picture Graph   111

Name _____   Date _____

Pretty Goldfish

Data Analysis and Probability

Jenna has 6 goldfish.
The circle graph shows what colors they are.
Each part of the graph stands for 1 goldfish.

Use the code to color the graph.
Then answer the questions.

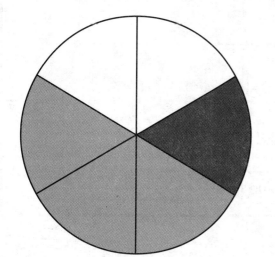

 red

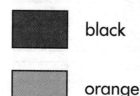 black

orange

A. How many goldfish are red? _____

B. How many goldfish are black? _____

C. How many goldfish are orange? _____

D. How many goldfish are either black or red? _____

E. How many goldfish are either orange or red? _____

F. How many more goldfish are orange than black? _____

Standards-Based Math • 1–2 © 2004 Creative Teaching Press

Name _____ Date _____

Timmy's Caps

Data Analysis and Probability

Timmy has 8 caps.
The circle graph shows what colors they are.
Each part of the circle stands for 1 cap.

Use the code to color the graph.
Then answer the questions.

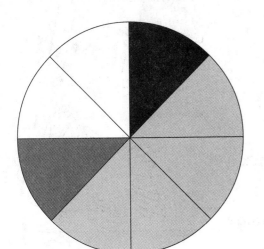

 white

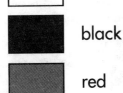 black

red

 blue

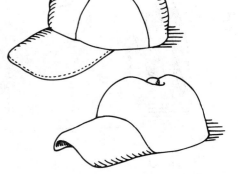

A. How many caps are black? _____

B. How many caps are blue? _____

C. Are there more white caps or red caps? _____

D. Are there more blue caps or white caps? _____

E. How many caps are either black or white? _____

F. How many caps are either blue or red? _____

G. Which cap color do you think Timmy likes the best?
Why? _____

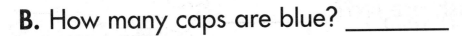

Name _____   Date _____

A Box of Buttons

Data Analysis and Probability

Use the code to color the buttons.

| p — purple | r — red | y — yellow |

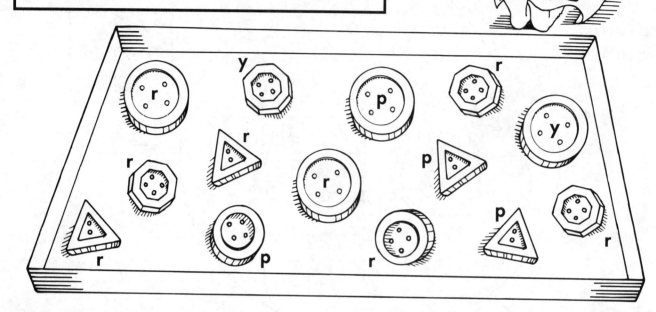

Pretend that you are going to close your eyes and pick up a button.
Answer the questions to tell what color you might pick up.

A. Is it more likely or less likely that
you will pick up a yellow button?   _____

B. Is it more likely or less likely that
you will pick up a red button?   _____

C. Do you have a better chance of
picking a red or purple button?   _____

D. Which color do you have the
least chance of picking up?   _____

Standards-Based Math • 1–2 © 2004 Creative Teaching Press

Name _____ Date _____

Gumball Picking

Data Analysis and Probability

Solve the problems. Use the answers to color the gumballs.
Then answer the questions at the bottom of the page.

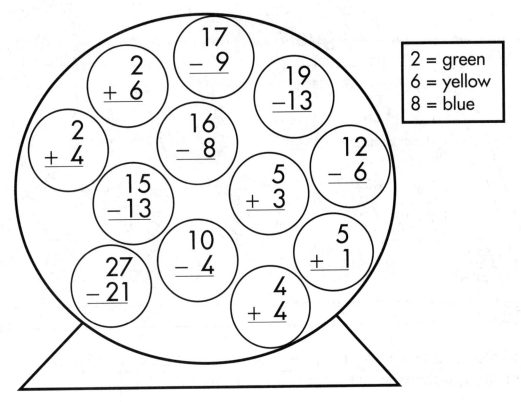

2 = green
6 = yellow
8 = blue

A. Is it more likely or less likely that you will get a green gumball? _____

B. Is it more likely or less likely that you will get a yellow gumball? _____

C. Do you have a better chance of getting a blue or yellow gumball? _____

D. Do you have a better chance of getting a blue or green gumball? _____

Who Will Win?

Data Analysis and Probability

Make a bar graph showing how many raffle tickets each child has.

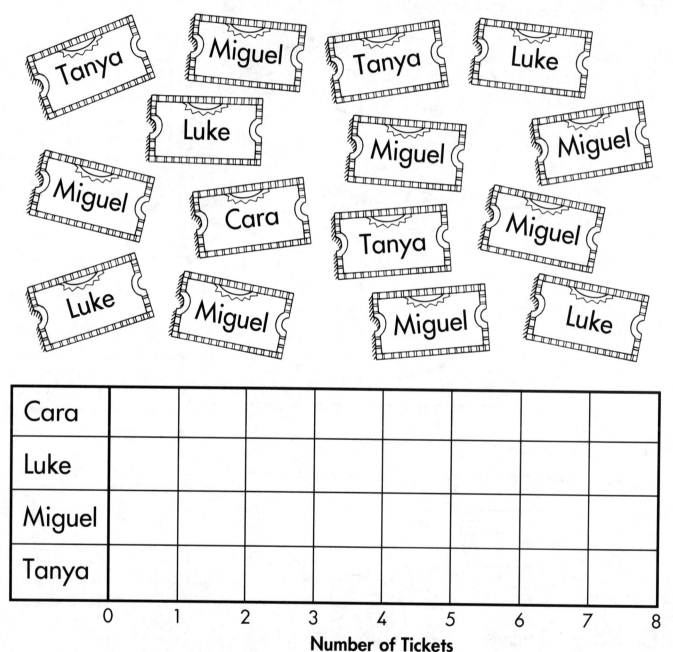

Number of Tickets

Who has the best chance of winning the raffle?

_____

Standards-Based Math • 1–2 © 2004 Creative Teaching Press

Name _____   Date _____

Potted Flowers

Data Analysis and Probability

Mrs. Lane has red, yellow, and purple flowers.
She has green, blue, and brown pots.

Mrs. Lane puts one flower in each pot.
Color the pictures to show the different ways
Mrs. Lane can mix and match her flowers and pots.

Answer Key

Batter Up! (page 5)

A. 3, 5, 9
B. 6, 3, 8
C. 10, 2, 8
D. 7, 9, 7
E. 10, 9, 9

Hop and Add (page 6)

A. 10, 11
B. 9, 12, 12
C. 11, 10, 9
D. 11, 12, 10
E. 11, 12, 12

Colorful Flowers (page 7)

A. 14, 13, 14, 15   (blue, red, blue, yellow)
B. 15, 16, 13, 16   (yellow, orange, red, orange)
C. 17, 15, 14, 13   (purple, yellow, blue, red)
D. 14, 18, 15, 16   (blue, green, yellow, orange)

Number Detective (page 8)

A. 4, 5
B. 5, 3, 2
C. 6, 9, 3
D. 5, 6, 6

E. 7, 4, 7
F. 9, 8, 9
G. 8, 7, 6
H. 8, 7, 8

Picnic Time (page 9)

	6 − 1 = _5_	5 − 4 = _1_	3 − 1 = _2_
5 − 3 = _2_	2 − 0 = _2_	7 − 4 = _3_	9 − 1 = _8_
8 − 4 = _4_	10 − 2 = _8_	8 − 3 = _5_	6 − 4 = _2_
10 − 5 = _5_	8 − 5 = _3_	9 − 4 = _5_	7 − 1 = _6_
6 − 3 = _3_	8 − 6 = _2_	7 − 3 = _4_	10 − 8 = _2_
9 − 2 = _7_	10 − 6 = _4_	8 − 2 = _6_	9 − 3 = _6_
10 − 3 = _7_	9 − 6 = _3_	10 − 1 = _9_	

Hop and Subtract (page 10)

A. 4, 7
B. 2, 2, 3
C. 7, 6, 3
D. 8, 5, 8
E. 6, 2, 2

Subtraction Roundup (page 11)

A. 5, 7, 8, 5, 6
B. 6, 4, 7, 6, 5
C. 8, 9, 9, 7, 8
D. 6, 8, 9, 7, 8
E. 9, 8, 9, 6, 9

Splish, Splash! (page 12)

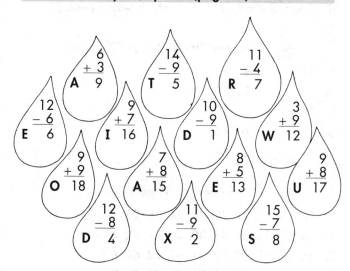

Answer to riddle: IT WEARS A DUXEDO!

Add It Up (page 13)

Sandra — 6; Yes   Sam — 8; Yes
Will — 3; No   Lyle — 5; No
Mackenzie — 4; No   Ashley — 7; Yes

Party Time (page 14)

A. 11   B. 15   C. 12   D. 14
E. 13   F. 11   G. 13
H. 14   I. 15   J. 18   K. 16
L. 13   M. 18   N. 17

Button Math (page 15)

A. 10 red, 4 yellow
1 ten and 4 ones = 14

B. 10 blue, 2 purple
1 ten and 2 ones = 12

C. 10 yellow, 5 green
1 ten and 5 ones = 15

D. 10 orange, 7 red
1 ten and 7 ones = 17

Bundles of Ten (page 16)

A. 4 tens, 40   **B.** 6 tens, 60   **C.** 2 tens, 20
D. 1 ten, 10   **E.** 3 tens, 30   **F.** 7 tens, 70
G. 9 tens, 90   **H.** 5 tens, 50   **I.** 8 tens, 80

Tens and Ones (page 17)

A. 3 tens, 5 ones; 35   **B.** 2 tens, 7 ones; 27
C. 6 tens, 1 one; 61   **D.** 5 tens, 4 ones; 54
E. 4 tens, 8 ones; 48   **F.** 7 tens, 6 ones; 76
G. 8 tens, 2 ones; 82   **H.** 5 tens, 5 ones; 55

Number Riddles (page 18)

A. 32   **B.** 16
C. 43   **D.** 89
E. 72   **F.** 29
G. 91   **H.** 45
I. 56   **J.** 61

All Aboard! (page 19)

A. 18, 19, 20, **21**, **22**, **23**, 24, **25**, **26**, 27
B. 34, 35, 36, **37**, **38**, **39**, 40, **41**, **42**, 43
C. 56, 57, 58, **59**, **60**, **61**, 62, **63**, **64**, 65
D. 65, 66, 67, **68**, **69**, **70**, 71, **72**, **73**, 74
E. 88, 89, **90**, **91**, **92**, 93, **94**, **95**, **96**, 97

F. **27**, 28, **29**   **G.** **70**, 71, **72**
H. **48**, 49, **50**   **I.** **89**, 90, **91**

Watch the Signs (page 20)

A. > < >   **D.** > < >
B. < < >   **E.** < > <
C. > < >   **F.** > > <

Rabbit's Riddle (page 21)

First row: 18, 68, 27, 48
Second row: 28, 38, 69, 77
Third row: 76, 98, 85, 59

Answer to riddle: A funny bunny!

Starry Subtraction (page 22)

A. 11, 15, 20
B. 10, 67, 84, 51
C. 41, 34, 50, 71
D. 95, 61, 80, 42
E. 73, 53, 91, 80

In the Clouds (page 23)

A. 64, 63
B. 13, 23, 78, 68, 78
C. 72, 12, 75, 53, 29
D. 63, 59, 5, 5, 66
E. 10, 85, 99, 75, 37

Eager Beavers (page 24)

A. 81, 82, 53, 95, 60
B. 31, 81, 95, 50, 35
C. 70, 92, 43, 53, 91
D. 93, 90, 82, 95, 80

A Cool Reminder (page 25)

A. 29, 7, 28, 31, 17   **C.** 54, 5, 18, 8, 65
B. 7, 8, 46, 4, 59   **D.** 85, 28, 29, 59, 47

Hundreds, Tens, and Ones (page 26)

A. 4 hundreds   2 tens   6 ones;   426
B. 1 hundred   7 tens   2 ones;   172
C. 3 hundreds   5 tens   1 one;   351
D. 6 hundreds   0 tens   5 ones;   605
E. 5 hundreds   4 tens   0 ones;   540

F. 8 hundreds   3 tens   7 ones
G. 2 hundreds   0 tens   9 ones
H. 7 hundreds   2 tens   0 ones

Zoomin' with Numbers (page 27)

A. 103, 104, **105**, **106**, 107, **108**, **109**, 110, **111**, 112
B. 345, 346, **347**, **348**, 349, **350**, **351**, 352, **353**, 354
C. 517, 518, **519**, **520**, **521**, 522, **523**, **524**, **525**, 526
D. 897, 898, **899**, **900**, **901**, 902, **903**, **904**, **905**, 906

E. 235, 325, 523, 535
F. 149, 194, 419, 491
G. 376, 673, 736, 763
H. 509, 590, 905, 950

A Space Trip (page 28)

A. 487	**F.** 760	**K.** 449	**P.** 795
B. 268	**G.** 449	**L.** 401	**Q.** 306
C. 562	**H.** 200	**M.** 110	**R.** 10
D. 900	**I.** 187	**N.** 200	
E. 721	**J.** 914	**O.** 399	

Wise Owl's Reminder (page 29)

A. 581, 482, 935, 977, 592
B. 560, 864, 649, 510, 810
C. 909, 527, 400, 724, 567
D. 898, 771, 982, 884, 680

Super Subtraction (page 30)

A. 657, 178, 211, 258, 164
B. 470, 99, 18, 391, 565
C. 7, 204, 209, 184, 337
D. 641, 449, 208, 694, 179

High-Flying Problems (page 31)

A. 631	**B.** 516	**C.** 507	**D.** 145
E. 268	**F.** 675	**G.** 108	
H. 7	**I.** 742	**J.** 904	**K.** 945
L. 306	**M.** 773	**N.** 590	

Even/Yellow: B, E, G, I, J, L, N
Odd/Orange: A, C, D, F, H, K, M

How Many Cents? (page 32)

A. 7¢	**B.** 16¢	**C.** 30¢
D. 25¢	**E.** 20¢	**F.** 20¢
G. 56¢	**H.** 42¢	**I.** 65¢

Counting Coins (page 33)

A. 28¢	**B.** 55¢	**C.** 35¢
D. 55¢	**E.** 80¢	**F.** 75¢
G. 50¢	**H.** 75¢	**I.** 66¢

Let's Go Shopping (page 34)

The following coins should be crossed out:

A. 1 dime and 1 nickel
B. 2 quarters
C. 1 quarter, 1 dime, and 1 nickel
D. 2 quarters and 1 nickel
E. 3 quarters
F. 3 quarters and 1 nickel
G. 2 quarters, 1 dime, and 2 nickels
H. 3 quarters, 1 dime, and 1 nickel

Zoo Tickets (page 35)

A. 3 tickets, 21¢ left
B. 2 tickets, 15¢ left
C. 3 tickets, 10¢ left
D. 2 tickets, 5¢ left
E. 2 tickets, 0¢ left
F. 1 ticket, 10¢ left

What Makes a Dollar? (page 36)

A. 100
B. 20
C. 10
D. 4
E. 2

Coin combinations will vary.

How Much Money? (page 37)

A. $1.35
B. $1.40
C. $2.50
D. $3.00
E. $2.75
F. $4.10
G. $3.61

Find the Fraction (page 38)

A. 1/3	**B.** 1/2	**C.** 1/4
D. 1/5	**E.** 1/3	**F.** 1/6
G. 1/4	**H.** 1/8	**I.** 1/6

A Closer Look at Fractions (page 39)

A. 2/3	**B.** 5/6	**C.** 1/2
D. 1/4	**E.** 3/5	**F.** 7/8
G. 2/5	**H.** 5/8	**I.** 3/4

Equal Groups (page 40)

The following number of items should be colored:

A. 2	**B.** 3
C. 4	**D.** 2
E. 4	**F.** 2

Fishy Fractions (page 41)

The following number of fish should be circled and colored:

A. 4	**B.** 2
C. 9	**D.** 3
E. 6	**F.** 4
G. 5	**H.** 8

Leafy Multiplication (page 42)

A. 10
B. 2
C. 16
D. 6
E. 20
F. 22

G. 4
H. 8
I. 14
J. 12
K. 24
L. 18

2, 4, 6, 8, 10, 12,
14, 16, 18, 20, 22, 24

Answers may vary. Possible answers include: The answers are even numbers. They are the same numbers you get when you count by twos.

Pretty Petals (page 43)

A. 5
B. 10
C. 15
D. 20
E. 25

F. 30
G. 35
H. 40
I. 45
J. 50

K. 10
L. 20
M. 30
N. 40
O. 50

P. 60
Q. 70
R. 80
S. 90
T. 100

Answers may vary. Possible answers include: The answers end in either 5 or 0. They are the same numbers you get when you count by fives or tens.

Dive for Treasure (page 44)

A. 4
B. 15
C. 20
D. 6
E. 50
F. 20

G. 2
H. 30
I. 10
J. 35
K. 16
L. 40

M. 30
N. 70
O. 45
P. 80
Q. 8

Bead Patterns (page 45)

Beads should be colored and the pattern continued as follows:

A. red, yellow, orange, red, yellow, orange
B. red, blue, blue, yellow
C. orange, yellow, orange, blue
D. Patterns will vary.

Shape Patterns (page 46)

Patterns should be completed as follows:

A.    D.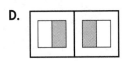

B.    E.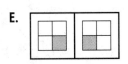

C. 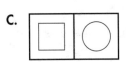   F.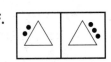

Circle Add-Ons (page 47)

Patterns should be completed as follows:

A.   C.

B.   D.

S-s-snaky Patterns (page 48)

A. 2, 4, **6**, **8**, 10, **12**, 14, 16
B. 18, 20, **22**, **24**, 26, **28**, **30**, 32
C. 36, 38, **40**, **42**, 44, **46**, **48**, 50
D. 54, 56, **58**, **60**, 62, 64, **66**, 68
E. 70, 72, **74**, **76**, 78, **80**, **82**, 84
F. 86, 88, **90**, **92**, **94**, 96, **98**, 100

Bubbly Numbers (page 49)

A. 10, 20, **30**, **40**, **50**, **60**, **70**, **80**, **90**, 100
B. 5, 10, **15**, **20**, **25**, **30**, 35, **40**, **45**, 50
C. 55, 60, **65**, **70**, **75**, 80, **85**, **90**, **95**, 100
D. 100, **90**, **80**, **70**, 60, **50**, **40**, **30**, **20**, 10
E. 85, 80, **75**, **70**, **65**, 60, **55**, **50**, **45**, 40

High in the Sky (page 50)

A. 3, 6, 9, 12, **15**, **18**, **21**, **24**
B. 7, 17, 27, 37, **47**, **57**, **67**, **77**
C. 30, 28, 26, 24, **22**, **20**, **18**, **16**
D. 95, 85, 75, 65, **55**, **45**, **35**, **25**
E. 1, 5, 9, 13, **17**, **21**, **25**, **29**
F. 10, 21, 32, 43, **54**, **65**, **76**, **87**

Fishy Facts (page 51)

A. 4 + 2 = 6; 2 + 4 = 6; 6 − 2 = 4; 6 − 4 = 2
B. 3 + 4 = 7; 4 + 3 = 7; 7 − 3 = 4; 7 − 4 = 3
C. 4 + 5 = 9; 5 + 4 = 9; 9 − 4 = 5; 9 − 5 = 4
D. 5 + 2 = 7; 2 + 5 = 7; 7 − 5 = 2; 7 − 2 = 5
E. 3 + 7 = 10; 7 + 3 = 10; 10 − 3 = 7; 10 − 7 = 3
F. 6 + 3 = 9; 3 + 6 = 9; 9 − 3 = 6; 9 − 6 = 3

Domino Math (page 52)

A. 2 + 8 = 10; 8 + 2 = 10; 10 − 2 = 8; 10 − 8 = 2
B. 3 + 6 = 9; 6 + 3 = 9; 9 − 3 = 6; 9 − 6 = 3
C. 7 + 5 = 12; 5 + 7 = 12; 12 − 7 = 5; 12 − 5 = 7
D. 9 + 2 = 11; 2 + 9 = 11; 11 − 9 = 2; 11 − 2 = 9
E. 3 + 5 = 8; 5 + 3 = 8; 8 − 3 = 5; 8 − 5 = 3
F. 4 + 8 = 12; 8 + 4 = 12; 12 − 4 = 8; 12 − 8 = 4
G. 3 + 8 = 11; 8 + 3 = 11; 11 − 3 = 8; 11 − 8 = 3
H. 9 + 3 = 12; 3 + 9 = 12; 12 − 9 = 3; 12 − 3 = 9
I. 5 + 6 = 11; 6 + 5 = 11; 11 − 5 = 6; 11 − 6 = 5

Number Houses (page 53)

A. 6 + 7 = 13; 7 + 6 = 13; 13 − 6 = 7; 13 − 7 = 6
B. 8 + 7 = 15; 7 + 8 = 15; 15 − 8 = 7; 15 − 7 = 8
C. 9 + 5 = 14; 5 + 9 = 14; 14 − 9 = 5; 14 − 5 = 9
D. 4 + 9 = 13; 9 + 4 = 13; 13 − 4 = 9; 13 − 9 = 4
E. 8 + 6 = 14; 6 + 8 = 14; 14 − 8 = 6; 14 − 6 = 8
F. 7 + 9 = 16; 9 + 7 = 16; 16 − 7 = 9; 16 − 9 = 7
G. 6 + 9 = 15; 9 + 6 = 15; 15 − 6 = 9; 15 − 9 = 6
H. 5 + 8 = 13; 8 + 5 = 13; 13 − 5 = 8; 13 − 8 = 5
I. 9 + 8 = 17; 8 + 9 = 17; 17 − 9 = 8; 17 − 8 = 9

Seesaw Sums (page 54)

A. 4 + 1; **5**   B. 4 + 4; **8**
C. 4 + 3; **7**   D. 2 + 2; **4**
E. 0 + 10; **10**   F. 6 + 3; **9**
G. 4 + 5; **9**   H. 2 + 4; **6**

Seesaw Subtraction (page 55)

A. 4 − 2; **2**   B. 3 − 1; **2**
C. 6 − 2; **4**   D. 9 − 2; **7**
E. 8 − 3; **5**   F. 7 − 1; **6**
G. 2 − 1; **1**   H. 5 − 2; **3**

Seesaw Challenge (page 56)

A. 10 + 3; **13**   B. 8 + 7; **15**
C. 2 + 7; **9**   D. 17 − 9; **8**
E. 12 − 4; **8**   F. 16 − 8; **8**
G. 14 − 7; **7**   H 15 − 8; **7**

Clowning Around (page 57)

A. 4 peanuts
B. 2 more balls
C. 12 parties
D. 7 clowns

Willie Wizard (page 58)

A. 8 mice
B. 5 lizards
C. 35 fish
D. 5 times

Draw and Add (page 59)

Pictures should be drawn and the problems solved to illustrate the following facts:

A. 4 + 5 = 9   5 + 4 = 9
B. 3 + 7 = 10   7 + 3 = 10
C. 8 + 6 = 14   6 + 8 = 14
D. 7 + 5 = 12   5 + 7 = 12

The sum does not change when you change the order of the numbers being added.

Happy Birthday! (page 60)

A. 6 boys, 6 girls
B. 8 gifts with bows, 4 gifts with no bows
C. 5 blue balloons, 10 red balloons
D. 5 yellow candles, 3 red candles

Piggy Bank Puzzlers (page 61)

A. 1 dime, 2 nickels

B. 1 quarter, 1 dime

C. 1 quarter, 2 dimes

D. 1 quarter, 1 dime, 1 nickel

E. 1 quarter, 1 dime, 2 nickels

F. 2 quarters, 2 nickels

G. 3 quarters, 1 nickel

H. 2 quarters, 1 dime, 2 nickels

Missing Signs (page 62)

A. $5 + 1 - 2 = 4$

B. $4 + 4 - 1 = 7$

C. $3 - 1 + 3 = 5$

D. $6 - 2 + 5 = 9$

E. $8 + 3 - 4 = 7$

F. $9 + 4 - 2 = 11$

G. $12 - 3 - 3 = 6$

H. $7 + 6 + 2 = 15$

Draw and Multiply (page 63)

Pictures should be drawn and the problems solved to illustrate the following facts:

A. $2 \times 4 = 8$; $4 \times 2 = 8$

B. $3 \times 5 = 15$; $5 \times 3 = 15$

C. $4 \times 3 = 12$; $3 \times 4 = 12$

D. $5 \times 2 = 2$; $2 \times 5 = 12$

The product does not change when you change the order of the numbers being multiplied.

Hidden Facts (page 64)

A. $5 \times 2 = 10$

B. $1 \times 5 = 5$

C. $2 \times 2 = 4$

D. $4 \times 10 = 40$

E. $2 \times 5 = 10$

F. $1 \times 2 = 2$

G. $10 \times 5 = 50$

H. $8 \times 5 = 40$

I. $4 \times 5 = 20$

J. $8 \times 10 = 80$

K. $3 \times 5 = 15$

L. $6 \times 2 = 12$

M. $10 \times 7 = 70$

N. $6 \times 5 = 30$

O. $5 \times 5 = 25$

P. $9 \times 5 = 45$

Q. $3 \times 2 = 6$

R. $8 \times 2 = 16$

S. $9 \times 2 = 18$

T. $6 \times 10 = 60$

Bagfuls of Numbers (page 65)

A. 41

B. 56

C. 38

D. 47

E. 62

F. 29

G. 63

H. 78

Evan's Riddles (page 66)

A. 42

B. 19

C. 27

D. 64

Sides and Corners (page 67)

A. 3 sides, 3 corners

B. 4 sides, 4 corners

C. 4 sides, 4 corners

D. 5 sides, 5 corners

E. 6 sides, 6 corners

F. 4 sides, 4 corners

G. 4 sides, 4 corners

H. 0 sides, 0 corners

A Shape Picture (page 68)

The picture should be colored accordingly.

There are 3 squares, 3 circles, 3 rectangles, and 6 triangles in the picture.

Dot-to-Dot Shapes (page 69)

The following shapes should be formed and named:

A. rectangle

B. triangle

C. hexagon

D. square

E. pentagon

F. octagon

Same Size and Shape (page 70)

A. **D.**

B. **E.**

C.

Sets of Shapes (page 71)

A. Sam put large shapes in Set 1 and small shapes in Set 2.

B. Pam put shapes with 4 sides in Set 1. She put shapes that did not have 4 sides in Set 2.

Pairs of Shapes (page 72)

A.

C.

B.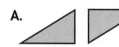

D.

Put Them Together (page 73)

A.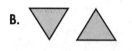

D.

B.

E.

C.

F.

Abracadabra! (page 74)

A.

B.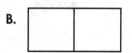

C.

D.

E and **F.** Answers will vary. Examples:

Tricky Squares (page 75)

A. 5   **B.** 5
C. 3   **D.** 7
E. 8   **F.** 7

Tricky Triangles (page 76)

A. 2   **B.** 3
C. 5   **D.** 6
E. 8   **F.** 5

Faces and Shapes (page 77)

The following shapes should be colored:
A. square   **B.** triangle
C. rectangle   **D.** square
E. circle   **F.** square

Blocks and Bases (page 78)

A. circle   **B.** square
C. rectangle   **D.** circle
E. square   **F.** rectangle

What Am I? (page 79)

A. pyramid   **B.** cube
C. rectangular prism   **D.** sphere
E. cylinder   **F.** cone

Exact Halves (page 80)

Lines should be drawn as shown.

A.

B.

C.

D.

E.

F.

G.

H.

I.

Kite Designs (page 81)

A.   B.

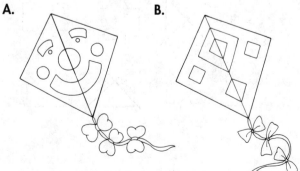

Paper Cutouts (page 82)

A.

B.

C.

D.

E.

F.

Which Shape Doesn't Belong? (page 83)

A.

B.

C.

D.

E.

F.

Slides, Flips, and Turns (page 84)

A.　　　　　D.

B.　　　　　E.

C.　　　　　F.

Find the Shape (page 85)

A.　　　　　B.

C.　　　　　D.

E.　　　　　F.

Where Is the Toy? (page 86)

A. C2　　**B.** D5
C. E1　　**D.** B1
E. A6　　**F.** E6
G. D3　　**H.** B4

What Time Is It? (page 87)

A. 4:00, 2:30, 6:00, 5:30
B. 7:30, 1:00, 5:00, 3:30
C. 11:00, 3:00, 4:30, 11:30
D. 12:00, 6:30, 12:30, 9:30

Feeding Time (page 88)

A. 6:00　　**B.** 6:30　　**C.** 7:00
D. 8:00　　**E.** 8:30　　**F.** 9:30
G. 10:00　　**H.** 10:30　　**I.** 11:30

125

Watch the Clock (page 89)

A. 7:45, quarter to 8
B. 2:45, quarter to 3
C. 5:15, quarter past 5
D. 4:15, quarter past 4
E. 6:15, quarter past 6
F. 12:45, quarter to 1
G. 3:15, quarter past 3
H. 4:45, quarter to 5
I. 3:45, quarter to 4

Passing Time (page 90)

A. 5:00
B. 2:30
C. 6:00
D. 7:15
E. 3:30
F. 1:45

Here Come the Months (page 91)

A. September
B. March
C. February
D. July

May Days (page 92)

A. Wednesday
B. 31
C. Tuesday
D. Sunday
E. 4
F. May 18
G. May 29
H. Wednesday, Thursday, Friday

Stringing Along (page 93)

A. 3
B. 1
C. 2
D. 4
E. 6
F. 5

Going to the Fair (page 94)

A. 3
B. 2
C. 1
D. 6
E. 3
F. 4
G. 2
H. 3

Centimeter Caterpillars (page 95)

A. 6
B. 4
C. 9
D. 8
E. 12
F. 15

Ounces and Pounds (page 96)

A. ounces
B. ounces
C. pounds
D. pounds
E. ounces
F. ounces
G. ounces
H. ounces
I. pounds

Grams and Kilograms (page 97)

A. kilograms
B. grams
C. kilograms
D. grams
E. kilograms
F. grams
G. grams
H. kilograms
I. grams

Cups and Pints (page 98)

A. 2
B. 4
C. 6
D. 8

Cups, Pints, and Quarts (page 99)

A. 2 pints
B. 5 cups
C. 2 quarts
D. 3 pints
E. 3 pints
F. 2 quarts

Equal Measures (page 100)

A. 2 cups
4 cups
6 cups
8 cups

B. 2 pints
4 cups
4 pints
8 cups

C. 4 quarts
8 quarts
12 quarts
16 quarts

D. 8 pints
16 cups
16 pints
32 cups

Liquid Measures (page 101)

A. 1 cup
B. 1 quart
C. 1 gallon
D. 1 pint
E. 1 gallon

Square Inches (page 102)

A. 2 **B.** 3
C. 4 **D.** 3
E. 5 **F.** 4

Find the Area (page 103)

A. 4 **B.** 1
C. 2 **D.** 3
E. 5 **F.** 4

Square Centimeters (page 104)

A. 8
B. 9
C. 15
D. 10
E. 5
F. 13
G. 8

Units of Measure (page 105)

A. inches
B. pounds
C. hours
D. ounces
E. cups
F. feet
G. minutes

Sky Gazing (page 106)

  7
  3
  1
  4

Show Time (page 107)

A. 7
B. balloons
C. 2

All Kinds of Weather (page 108)

A. 10
B. 4
C. 3
D. 13
E. 5

A Color Graph (page 109)

A. 5
B. 6
C. green
D. blue
E. red
F. 3

A Fruity Graph (page 110)

A. 5
B. 2
C. apple
D. bananas
E. 13
F. 24

Zoo Animals (page 111)

A. 4
B. 12
C. penguins
D. 10
E. 4
F. 2

Pretty Goldfish (page 112)

A. 2
B. 1
C. 3
D. 3
E. 5
F. 2

Timmy's Caps (page 113)

A. 1
B. 4
C. white
D. blue
E. 3
F. 5
G. blue; Timmy has more blue caps than any other color.

A Box of Buttons (page 114)

A. less likely
B. more likely
C. red
D. yellow

Gumball Picking (page 115)

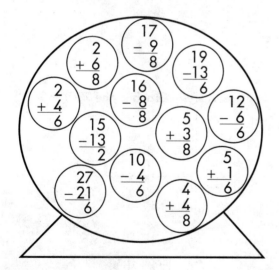

A. less likely
B. more likely
C. yellow
D. blue

Who Will Win? (page 116)

The bar graph should be filled in showing that Cara has 1 ticket, Luke has 4, Miguel has 7, and Tanya has 3.

Miguel has the best chance of winning the raffle.

Potted Flowers (page 117)

There are nine combinations: red flower/green pot; red flower/blue pot; red flower/brown pot; yellow flower/green pot; yellow flower/blue pot; yellow flower/brown pot; purple flower/green pot; purple flower/blue pot; purple flower/brown pot.